TEACHER RESOURCES

Electromagnetic Force

FOSS

Next Generation

Full Option Science System
Developed at the Lawrence Hall of Science, University of California, Berkeley
Published and Distributed by Delta Education

FOSS Lawrence Hall of Science Team
Larry Malone and Linda De Lucchi, FOSS Project Codirectors and Lead Developers; Jessica Penchos, Middle School Coordinator; Kathy Long, FOSS Assessment Director; David Lippman, Program Manager; Carol Sevilla, Publications Design Coordinator; Susan Stanley, Graphic Production; Rose Craig, Illustrator
FOSS Curriculum Developers: Alan Gould, Teri Lawson, Ann Moriarty, Virginia Reid, Joanna Snyder
FOSS Technology Developers: Susan Ketchner, Arzu Orgad
FOSS Multimedia Team: Kate Jordan, Senior Multimedia; Christopher Keller, Multimedia Producer; Jonathan Segal, Designer; Christopher Cianciarulo, Designer; Dan Bluestein, Programmer; Shan Jiang, Programmer

Delta Education Team
Bonnie A. Piotrowski, Editorial Director, Elementary Science
Project Team: Jennifer Apt, Sandra Burke, Mary Connell, Joann Hoy, Angela Miccinello, Jennifer Staltare

Content Reviewer
Bryan J. Mendez
Astronomer & Public Education Specialist, Space Sciences Laboratory
University of California, Berkeley

Thank you to all FOSS Middle School Revision Trial Teachers and District Coordinators
Jonah Cohn, St. Anthony's School, Oakland, CA; Frances Amojioyi, Lincoln Middle School, Alameda, CA; Dean Anderson, Organized trials for Boston Public Schools, Boston, MA; Thomas Archer, Organized trials for ESD 112, Vancouver, WA; Lauresa Baker, Lincoln Middle School, Alameda, CA; Bobbi Anne Barnowsky, Canyon Middle School, Castro Valley, CA; Christine Bertko, St. Finn Barr Catholic School, San Francisco, CA; Stephanie Billinge, James P. Timilty Middle School, Roxbury, MA; Jerry Breton, Ingleside Middle School, Phoenix, AZ; Robert Cho, Timilty Middle School, Boston, MA; Susan Cohen, Cherokee Heights Middle School, Madison, WI; Malcolm Davis, Canyon Middle School, Castro Valley, CA; Marilyn Decker, Organized trials for Milton PS, Milton, MA; Jenny Ernst, Park Day School, Oakland, CA; Marianne Floyd, Spanaway Middle School, Spanaway, WA; Sarah Kathryn Gessford, Journeys School, Jackson, WY; Charles Hardin, Prairie Point Middle School, Cedar Rapids, IA; Jennifer Hartigan, Lincoln Middle School, Alameda, CA; Sheila Holland, TechBoston Academy, Boston, MA; Nicole Hoyceanyls, Charles S. Pierce Middle School, Milton, MA; Bruce Kamerer, Donald McKay K-8 School, East Boston, MA; Carmen Saele Kardokus, Reeves Middle School, Olympia, WA; Janey Kaufman, Organized trials for Scottsdale USD, Scottsdale, AZ; Erica Larson, Organized trials for Grant Wood AEA, Cedar Rapids, IA; Lindsay Lodholz, O'Keeffe Middle School, Madison, WI; Robert Mattisinko, Chaparral High School, Scottsdale, AZ; Brenda McGurk, Prairie Point Middle School, Cedar Rapids, IA; Tim Miller, Mountainside Middle School, Scottsdale, AZ; Thomas Miro, Lincoln Middle School, Alameda, CA; Spencer Nedved, Frontier Middle School, Vancouver, WA; Joslyn Olsen, Lincoln Middle School, Alameda, CA; Stephanie Ovechka, Cedarcrest Middle School, Spanaway, WA; Barbara Reinert, Copper Ridge School, Scottsdale, AZ; Stephen Ramos, Lincoln Middle School, Alameda, CA; Gina Rutenbeck, Prairie Point Middle School, Cedar Rapids, IA; John Sheridan, Boston Public Schools (Boston Schoolyard Initiative), Boston, MA; Barbara Simon, Timilty Middle School, Boston, MA; Lise Simpson, Alcott Middle School, Norman, OK; Autumn Stevick, Thurgood Marshall Middle School, Olympia, WA; Ted Stoeckley, Hall Middle School, Larkspur, CA; Lesli Taschwer, Organized trials for Madison SD, Madison, WI; Paula Warner, Alcott Middle School, Norman, OK; Darren T. Wells, James P. Timilty Middle School, Boston, MA; Kristin White, Frontier Middle School, Vancouver, WA

Photo Credits: © iStockphoto/Bosca78 (cover); © Delta Education

Published and Distributed by Delta Education, a member of the School Specialty Family
The FOSS program was developed in part with the support of the National Science Foundation grant nos. ESI-9553600 and ESI-0242510. However, any opinions, findings, conclusions, statements, and recommendations expressed herein are those of the authors and do not necessarily reflect the views of NSF. FOSSmap was developed in collaboration between the BEAR Center at UC Berkeley and FOSS at the Lawrence Hall of Science. Score analysis is done through the BEAR Center Scoring Engine.

Electromagnetic Force — Teacher Toolkit, 1465641
Teacher Resources, 1465692
978-1-62571-205-9
Printing 1 – 9/2016
Patterson Printing, Benton Harbor, MI

TABLE OF CONTENTS

This document, *Teacher Resources*, is one of three parts of the *FOSS Teacher Toolkit* for this course. The chapters in *Teacher Resources* are all available as PDFs on FOSSweb.

The other parts of the course *Teacher Toolkit* are the *Investigations Guide* and a copy of the *FOSS Science Resources* student book containing original readings for this course.

The spiral-bound *Investigations Guide* contains these chapters.

• Overview
• Framework and NGSS
• Materials
• Investigations
• Assessment

The *Teacher Toolkit* is the most important part of the FOSS Program. It is here that all the wisdom and experience contributed by hundreds of educators has been assembled. Everything we know about the content of the course, how to teach the subject, and the resources that will assist the effort are presented here.

FOSS Program Goals

Contents

INTRODUCTION

The Full Option Science System™ has evolved from a philosophy of teaching and learning at the Lawrence Hall of Science that has guided the development of successful active-learning science curricula for more than 40 years. The FOSS Program bridges research and practice by providing tools and strategies to engage students and teachers in enduring experiences that lead to deeper understanding of the natural and designed worlds.

Science is a creative and analytic enterprise, made active by our human capacity to think. Scientific knowledge advances when scientists observe objects and events, think about how they relate to what is known, test their ideas in logical ways, and generate explanations that integrate the new information into understanding of the natural world. Engineers apply that understanding to solve real-world problems. Thus the scientific enterprise is both what we know (content knowledge) and how we come to know it (practices). Science is a discovery activity, a process for producing new knowledge.

The best way for students to appreciate the scientific enterprise, learn important scientific and engineering concepts, and develop the ability to think well is to actively participate in scientific practices through their own investigations and analyses. FOSS was created to engage students and teachers with meaningful experiences in the natural and designed worlds.

GOALS OF THE FOSS PROGRAM

FOSS has set out to achieve three important goals: scientific literacy, instructional efficiency, and systemic reform.

Scientific Literacy

FOSS provides all students with science experiences that are appropriate to students' cognitive development and prior experiences. It provides a foundation for more advanced understanding of core science ideas that are organized in thoughtfully designed learning progressions and prepares students for life in an increasingly complex scientific and technological world.

The National Research Council (NRC) in *A Framework for K–12 Science Education: Practices, Crosscutting Concepts, and Core Ideas* and the American Association for the Advancement of Science (AAAS) in *Benchmarks for Scientific Literacy* have described the characteristics of scientific literacy:

- Familiarity with the natural world, its diversity, and its interdependence.

- Understanding the disciplinary core ideas and the crosscutting concepts of science, such as patterns; cause and effect; scale, proportion, and quantity; systems and system models; energy and matter—flows, cycles, and conservation; structure and function; and stability and change.

- Knowing that science and engineering, technology, and mathematics are interdependent human enterprises and, as such, have implied strengths and limitations.

- Ability to reason scientifically.

- Using scientific knowledge and scientific and engineering practices for personal and social purposes.

The FOSS Program design is based on learning progressions that provide students with opportunities to investigate core ideas in science in increasingly complex ways over time. FOSS starts with the intuitive ideas that primary students bring with them and provides experiences that allow students to develop more sophisticated understanding as they grow through the grades. Cognitive research tells us that learning involves individuals in actively constructing schemata to organize new information and to relate and incorporate the new understanding into established knowledge. What sets experts apart from novices is that experts in a discipline have extensive knowledge that is effectively organized into structured schemata to promote thinking. Novices have disconnected ideas about a topic that are difficult to retrieve and use. Through internal processes to establish schemata and through social processes of interacting with peers and adults, students construct understanding of the natural world and their relationship to it.

The target goal for FOSS students is to know and use scientific explanations of the natural world and the designed world; to understand the nature and development of scientific knowledge and technological capabilities; and to participate productively in scientific and engineering practices.

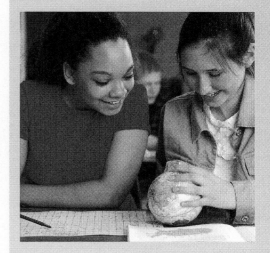

Instructional Efficiency

FOSS provides all teachers with a complete, cohesive, flexible, easy-to-use science program that reflects current research on teaching and learning, including student discourse, argumentation, writing to learn, and reflective thinking, as well as teacher use of formative assessment to guide instruction. The FOSS Program uses effective instructional methodologies, including active learning, scientific practices, focus questions to guide inquiry, working in collaborative groups, multisensory strategies, integration of literacy, appropriate use of digital technologies, and making connections to students' lives.

FOSS is designed to make active learning in science engaging for teachers as well as for students. It includes these supports for teachers:

- Complete equipment kits with durable, well-designed materials for all students.

- Detailed *Investigations Guide* with science background for the teacher and focus questions to guide instructional practice and student thinking.

- Multiple strategies for formative assessment.

- Benchmark assessments with scoring guides.

- Strategies for use of science notebooks for novice and experienced users.

- *FOSS Science Resources*, a book of course-specific readings.

- The FOSS website with course-integrated online activities for use in school or at home, suggested extension activities, and extensive online support for teachers.

Systemic Reform

FOSS provides schools and school systems with a program that addresses the community science-achievement standards. The FOSS Program prepares students by helping them acquire the knowledge and thinking capacity appropriate for world citizens.

The FOSS Program design makes it appropriate for reform efforts on all scales. It reflects the core ideas to be incorporated into the next-generation science standards. It meets with the approval of science and technology companies working in collaboration with school systems, and it has demonstrated its effectiveness with diverse student and teacher populations in major urban reform efforts. The use of science notebooks and formative-assessment strategies in FOSS redefines the role of science in a school—the way that teachers engage in science teaching with one another as professionals and with students as learners, and the way that students engage in science learning with the teacher and with one another. FOSS takes students and teachers beyond the classroom walls to establish larger communities of learners.

BRIDGING RESEARCH INTO PRACTICE

The FOSS Program is built on the assumptions that understanding core scientific knowledge and how science functions is essential for citizenship, that all teachers can teach science, and that all students can learn science. The guiding principles of the FOSS design, described below, are derived from research and confirmed through FOSS developers' extensive experience with teachers and students in typical American classrooms.

Understanding of science develops over time. FOSS has elaborated learning or content progressions for core ideas in science for kindergarten through grade 8. Developing the learning progressions involves identifying successively more sophisticated ways of thinking about core ideas over multiple years. "If mastery of a core idea in a science discipline is the ultimate educational destination, then well-designed learning progressions provide a map of the routes that can be taken to reach that destination" (National Research Council, *A Framework for K–12 Science Education*, 2011).

Focusing on a limited number of topics in science avoids shallow coverage and provides more time to explore core science ideas in depth. Research emphasizes that fewer topics experienced in greater depth produces much better learning than many topics briefly visited. FOSS affirms this research. FOSS courses provide long-term engagement (10–12 weeks) with important science ideas. Furthermore, courses build upon one another within and across each strand, progressively moving students toward the grand ideas of science. The core ideas of science are difficult and complex, never learned in one lesson or in one class year.

FOSS Next Generation—K–8 Sequence

		PHYSICAL SCIENCE		EARTH SCIENCE		LIFE SCIENCE	
		MATTER	ENERGY AND CHANGE	ATMOSPHERE AND EARTH	ROCKS AND LANDFORMS	STRUCTURE/ FUNCTION	COMPLEX SYSTEMS
8	6–8	Waves; Gravity and Kinetic Energy Chemical Interactions Electromagnetic Force; Variables and Design		Planetary Science Earth History Weather and Water		Heredity and Adaptation; Human Systems Interactions Populations and Ecosystems Diversity of Life	
	5	Mixtures and Solutions		Earth and Sun		Living Systems	
	4		Energy		Soils, Rocks, and Landforms	Environments	
	3	Motion and Matter		Water and Climate		Structures of Life	
	2	Solids and Liquids			Pebbles, Sand, and Silt	Insects and Plants	
	1		Sound and Light	Air and Weather		Plants and Animals	
K	K	Materials and Motion		Trees and Weather		Animals Two by Two	

FOSS Program Goals

Science is more than a body of knowledge. How well you think is often more important than how much you know. In addition to the science content framework, every FOSS course provides opportunities for students to engage in and understand science practices, and many courses explore issues related to engineering practices and the use of natural resources. FOSS uses these science and engineering practices.

- Asking questions (for science) and defining problems (for engineering)

- Developing and using models

- Planning and carrying out investigations

- Analyzing and interpreting data

- Using mathematics, information and computer technology, and computational thinking

- Constructing explanations (for science) and designing solutions (for engineering)

- Engaging in argument from evidence

- Obtaining, evaluating, and communicating information

Science is inherently interesting, and children are natural investigators. It is widely accepted that children learn science concepts best by doing science. Doing science means hands-on experiences with objects, organisms, and systems. Hands-on activities are motivating for students, and they stimulate inquiry and curiosity. For these reasons, FOSS is committed to providing the best possible materials and the most effective procedures for deeply engaging students with scientific concepts. FOSS students at all grade levels investigate, experiment, gather data, organize results, and draw conclusions based on their own actions. The information gathered in such activities enhances the development of science and engineering practices.

Education is an adventure in self-discovery. Science provides the opportunity to connect to students' interests and experiences. Prior experiences and individual learning styles are important considerations for developing understanding. Observing is often equated with seeing, but in the FOSS Program all senses are used to promote greater understanding. FOSS evolved from pioneering work done in the 1970s with students with disabilities. The legacy of that work is that FOSS investigations naturally use multisensory methods to accommodate students with physical and learning disabilities and also to maximize information gathering for all students. A number of tools, such as the FOSS syringe and balance, were originally designed to serve the needs of students with disabilities.

Formative assessment is a powerful tool to promote learning and can change the culture of the learning environment. Formative assessment in FOSS creates a community of reflective practice. Teachers and students make up the community and establish norms of mutual support, trust, respect, and collaboration. The goal of the community is that everyone will demonstrate progress and will learn and grow.

Science-centered language development promotes learning in all areas. Effective use of science notebooks can promote reflective thinking and contribute to lifelong learning. Research has shown that when language-arts experiences are embedded within the context of learning science, students improve in their ability to use their language skills. Students are motivated to read to find out information, and to share their experiences both verbally and in writing.

Experiences out of the classroom develop awareness of community. By extending classroom learning into the local region and community, FOSS brings the science concepts and principles to life. In the process of extending classroom learning to the natural world and utilizing community resources, students will develop a relationship with learning that extends beyond the classroom walls.

FOSS NEXT GENERATION K–8 SCOPE AND SEQUENCE

Grade	Physical Science	Earth Science	Life Science
6–8	Waves* Gravity and Kinetic Energy*	Planetary Science	Heredity and Adaptation* Human Systems Interactions*
	Chemical Interactions	Earth History	Populations and Ecosystems
	Electromagnetic Force* Variables and Design*	Weather and Water	Diversity of Life
5	Mixtures and Solutions	Earth and Sun	Living Systems
4	Energy	Soils, Rocks, and Landforms	Environments
3	Motion and Matter	Water and Climate	Structures of Life
2	Solids and Liquids	Pebbles, Sand, and Silt	Insects and Plants
1	Sound and Light	Air and Weather	Plants and Animals
K	Materials and Motion	Trees and Weather	Animals Two by Two

* Half-length course

FOSS is a research–based science curriculum for grades K–8 developed at the Lawrence Hall of Science, University of California, Berkeley. FOSS is also an ongoing research project dedicated to improving the learning and teaching of science. The FOSS project began over 25 years ago during a time of growing concern that our nation was not providing young students with an adequate science education. The FOSS Program materials are designed to meet the challenge of providing meaningful science education for all students in diverse American classrooms and to prepare them for life in the 21st century. Development of the FOSS Program was, and continues to be, guided by advances in the understanding of how people think and learn.

With the initial support of the National Science Foundation and continued support from the University of California, Berkeley, and School Specialty, Inc., the FOSS Program has evolved into a curriculum for all students and their teachers, grades K–8. The current editions of FOSS are the result of a rich collaboration among the FOSS/Lawrence Hall of Science development staff; the FOSS product development team at School Specialty; assessment specialists, educational researchers, and scientists; and dedicated professionals in the classroom and their students, administrators, and families.

We acknowledge the thousands of FOSS educators who have embraced the notion that science is an active process, and we thank them for their significant contributions to the development and implementation of the FOSS Program.

Science Notebooks
in Middle School

Source: Special Collections Research Center, University of Chicago Library

A scientist's notebook

A student's notebook

Contents

INTRODUCTION

Scientists keep notebooks. The scientist's notebook is a detailed record of his or her engagement with scientific phenomena. It is a personal representation of experiences, observations, and thinking—an integral part of the process of doing scientific work. A scientist's notebook is a continuously updated history of the development of scientific knowledge and reasoning. The notebook organizes the huge body of knowledge and makes it easier for a scientist to work. As developing scientists, FOSS students are encouraged to incorporate notebooks into their science learning. First and foremost, the notebook is a tool for student learning.

NOTEBOOK BENEFITS

Engaging in active science is one part experience and two parts making sense of the experience. Science notebooks help students with the sense-making part by providing two major benefits: documentation and cognitive engagement.

Benefits to Students

Science notebooks centralize students' data. When data are displayed in functional ways, students can think about the data more effectively. A well-kept notebook is a useful reference document. When students have forgotten a fact or relationship that they learned earlier in their studies, they can look it up. Learning to reference previous discoveries and knowledge structures is important.

Documentation: an organized record. As students become more accomplished at keeping notebooks, their work will become better organized and efficient. Tables, graphs, charts, drawings, and labeled illustrations will become standard means for representing and displaying data. A complete and accurate record of learning allows students to reconstruct the sequence of learning events and relive the experience. Discussions about science among students, students and teachers, or students, teachers, and families, have more meaning when they are supported by authentic documentation in students' notebooks. Questions and ideas generated by experimentation or discussion can be recorded for future investigation.

From the Human Systems Interactions Course

From the Weather and Water Course

Cognitive engagement. Once data are recorded and organized in an efficient manner in science notebooks, students can think about the data and draw conclusions about the way the world works. Their data are the raw materials that students use to forge concepts and relationships from their experiences and observations.

Writing stimulates active reasoning. There is a direct relationship between the formation of concepts and the rigors of expressing them in words. Writing requires students to impose discipline on their thoughts. When you ask students to generate derivative products (summary reports, detailed explanations, posters, oral presentations, etc.) as evidence of learning, the process will be much more efficient and meaningful because they have a coherent, detailed notebook for reference.

When students use notebooks as an integral part of their science studies, they think critically about their thinking. This reflective thinking can be encouraged by notebook entries that present opportunities for self-assessment. Self-assessment motivates students to rethink and restate their scientific understanding. Revising their notebook entries helps students clarify their understanding of the science concepts under investigation. By writing explanations, students clarify what they know and expose what they don't know.

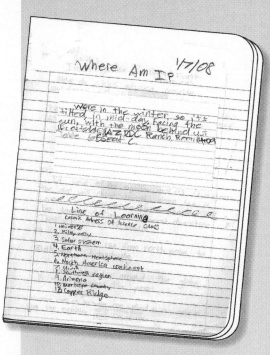

From the Planetary Science Course

From the Planetary Science Course

Benefits to Teachers

In FOSS, the unit of instruction is the course—a sequence of conceptually related learning experiences that leads to a set of learning outcomes. A science notebook helps you think about and communicate the conceptual structure of the course you are teaching.

Assessment. From the assessment point of view, a science notebook is a collection of student-generated artifacts that exhibit learning. You can informally assess student skills, such as using charts to record data, in real time while students are working with materials. At other times, you might collect student work samples and review them for insights or errors in conceptual understanding. This valuable information helps you plan the next steps of instruction. Students' data analysis, sense making, and reflection provide a measure of the quality and quantity of student learning. The notebook itself should not be graded, though certain assignments might be graded and placed in the notebook.

Medium for feedback. The science notebook provides an excellent medium for providing feedback to individual students regarding their work. Productive feedback calls for students to read a teacher comment, think about the issue it raises, and act on it. The comment may ask for clarification, an example, additional information, precise vocabulary, or a review of previous work in the notebook. In this way, you can determine whether a problem with the student work relates to a flawed understanding of the science content or a breakdown in communication skills.

Focus for professional discussions. The student notebook also acts as a focal point for discussion about student learning at several levels. First, a student's work can be the subject of a conversation between you and the student. By acting as a critical mentor, you can call attention to ways a student can improve the notebook, and help him or her learn how to use the notebook as a reference. You can also review and discuss the science notebook during family conferences. Science notebooks shared among teachers in a study group or other professional-development environment can effectively demonstrate recording techniques, individual styles, various levels of work quality, and so on. Just as students can learn notebook strategies from one another, teachers can learn notebook skills from one another.

GETTING STARTED

A middle school science notebook is more than just a collection of science work, notes, field-trip permission slips, and all the other types of documents that tend to accumulate in a student's three-ring binder or backpack. By organizing the science work systematically into a bound composition book, students create a thematic record of their experiences, thoughts, plans, reflections, and questions as they work through a topic in science.

The science notebook is more than just formal lab reports; it is a record of a student's entire journey through a progression of science concepts. Where elementary school students typically need additional help structuring and organizing their written work, middle school students should be encouraged to develop their organizational skills and take some ownership in creating deliberate records of their science learning, even though they may still require some pointers and specific scaffolding from you.

In addition, the science notebook provides a personal space where students can explore their understanding of science concepts by writing down ideas and being allowed to "mess around" with their thinking. Students are encouraged to look back on their ideas throughout the course to self-assess their conceptual development and record new thoughts. With this purpose of the science notebook in mind, you may need to refine your own thinking around what should or should not be included as a part of the science notebook, as well as expectations about grading and analyzing student work.

Rules of Engagement

Teachers and students should be clear about the conventions students will honor in their notebook entries. Typically, the rules of grammar and spelling are fairly relaxed so as not to inhibit the flow of expression during notebook entries. This also helps students develop a sense of ownership in their notebooks, a place where they are free to write in their own style. When students generate derivative products using information in the notebooks, such as reports, you might require students to exercise more rigorous language-arts conventions.

In addition to written entries, students should be encouraged to use a wide range of other means for recording and communicating, including charts, tables, graphs, drawings, graphics, color codes, numbers, and artifacts attached to the notebook pages. By expanding the options for making notebook entries, each student will find his or her most efficient, expressive way to capture and organize information for later retrieval.

Enhanced Classroom Discussion

One of the benefits of using notebooks is that you will elicit responses to key discussion questions from all students, not just the handful of verbally enthusiastic students in the class. When you ask students to write down their thoughts after you pose a question, all students have time to engage deeply with the question and organize their thoughts. When you ask students to share their answers, those who needed more time to process the question and organize their thinking will be ready to verbalize their responses and become involved in a class discussion.

When students can use their notebooks as a reference during the ensuing discussion, they won't feel put on the spot. At some points, you might ask students to share only what they wrote in their notebooks, to remind them to focus their thoughts while writing. As the class shares ideas during discussions, students can add new ideas to their notebooks under a line of learning (see next-step strategies). Even if some students are still reticent, having students write after a question is posed prevents them from automatically disengaging from conversations.

Notebook Structure

FOSS recommends that students keep their notebooks in 8" × 10" bound composition books. At the most advanced level, students are responsible for creating the entire science notebook from blank pages in their composition books. Experienced students determine when to use their notebooks, how to organize space, what methods of documentation to use, and how to flag important information. This level of notebook use will not be realized quickly; it will likely require systematic development by an entire teaching staff over time.

At the beginning, notebook practice is often highly structured, using prepared sheets from the FOSS notebook masters. You can photocopy and distribute these sheets to students as needed during the investigations. Sheets are sized to fit in a standard composition book. Students glue or tape the sheets into their notebooks. This allows some flexibility between glued-in notebook sheets and blank pages where students can do additional writing, drawings, and other documentation. Prepared notebook sheets are helpful organizers for students with challenges such as learning disabilities or with developing English skills. This model is the most efficient means for obtaining the most productive work from inexperienced middle school students.

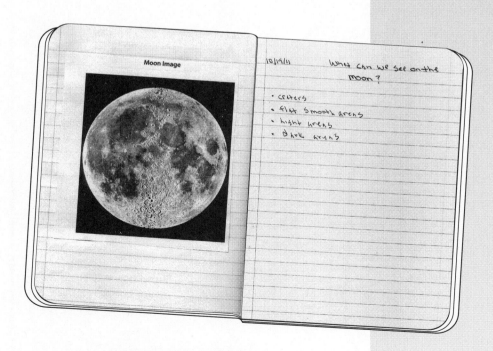

From the Planetary Science Course

To make it easy for new FOSS teachers to implement a beginning student notebook, Delta Education sells copies of the printed *FOSS Science Notebook* in English for all FOSS middle school courses. Electronic versions of the notebook sheets can be downloaded free of charge at www.FOSSweb.com.

Each *FOSS Science Notebook* is a bound set of the notebook sheets for the course plus extra blank sheets throughout the notebook for students to write focus or inquiry questions, record and organize data, make sense of their thinking, and write summaries. There are also blank pages at the end to develop an index of science vocabulary.

The questions, statements, and graphic organizers on the notebook sheets provide guidance for students and scaffolding for teachers. When the notebook sheets are organized as a series, they constitute a highly structured precursor to an autonomously generated science notebook.

Developing Notebook Skills

Students will initially need more guidance from you. You will need to describe what and when to record, and to model organizational techniques. As the year advances, the notebook work will become increasingly student centered. As the body of work in the notebook grows, students will have more and more examples of useful techniques for reference. This self-sufficiency reduces the amount of guidance you need to provide, and reinforces students' appreciation of their own record of learning.

This gradual shift toward student-centered use of the notebook applies to any number of notebook skills, including developing headers for each page (day, time, date, title, etc.); using space efficiently on the page; preparing graphs, graphic organizers, and labeled illustrations; and attaching artifacts (sand samples, dried flowers, photographs, etc.). For instance, when students first display their data in a two-coordinate graph, the graph might be completely set up for them, so that they simply plot the data. As the year progresses, they will be expected to produce graphs with less and less support, until they are doing so without any assistance from you.

Organizing Science Notebooks

Four organizational components of the notebook should be planned right from the outset: a table of contents, page numbering, entry format, and an index.

Table of contents. Students should reserve the first three to five pages of their notebook for the table of contents. They will add to it systematically as they proceed through the course. The table of contents should include the date, title, and page number for each entry. The title could be based on the names of the investigations in the course, the specific activities undertaken, the concepts learned, a focus question for each investigation, or some other schema that makes sense to everyone.

Page numbering. Each page should have a number. These are referenced in the table of contents as the notebook progresses.

Entry format. During each class session, students will document their learning. Certain information will appear in every record, such as the date and title. Other forms of documentation will vary, including different types of written entries and artifacts, such as a multimedia printout. Some teachers ask their students to start each new entry at the top of the next available page. Others simply leave a modest space before a new entry. Sometimes it is necessary to leave space for work that will be completed on a separate piece of paper and glued or taped in later. Students might also leave space after a response, so that they can add to it at a later time.

Index. Scientific academic language is important. FOSS strives to have students use precise, accurate vocabulary at all times in their writing and conversations. To help them learn scientific vocabulary, students should set up an index at the end of their notebooks. It is not usually possible for students to enter the words in alphabetical order, as they will be acquired as the course advances. Instead, students could use several pages at the end of the notebook blocked out in 24 squares, and assign one or more letters to each square. Students write the new vocabulary word or phrase in the appropriate square and tag it with the page number of the notebook on which the word is defined. By developing vocabulary in context, students construct meaning through the inquiry process, and by organizing the words in an index, they strengthen their science notebooks as a documentary tool of their science learning. As another alternative, students can also define the word within these squares with the page references.

Table of contents

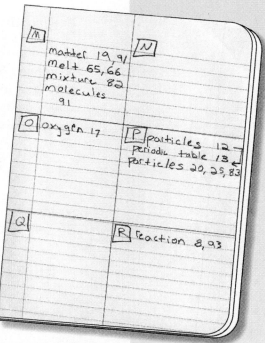

A science notebook index

NOTEBOOK COMPONENTS

Four general types of notebook entries, or components, give the science notebook conceptual shape and direction. These structures don't prescribe a step-by-step procedure for how to prepare the notebook, but they do provide some overall guidance. The general arc of an investigation starts with a question or challenge, proceeds with an activity and data acquisition, continues to sense making, and ends with next steps such as reflection and self-assessment.

All four components are not necessary during each class session, but over the course of an investigation, each component will be visited at least once. It may be useful to keep these four components in mind as you systematically guide students through their notebook entries. The components are overviewed here and described in greater detail on the following pages.

Focusing the investigation. Each part of each FOSS investigation includes a focus question, which students transcribe into their notebooks. Focus questions are embedded in the teacher step-by-step instructions and explicitly labeled. The focus question establishes the direction and conceptual challenge for that part of the investigation. For instance, when students investigate the origins of sand and sandstone in the **Earth History Course**, they start by writing,

> ➤ *Which came first, sand or sandstone?*

The question focuses both students and you on the learning goals for the activity. Students may start by formulating a plan, formally or informally, for answering the focus question. The goal of the plan is to obtain a satisfactory answer to the focus question, which will be revisited and answered later in the investigation.

Data acquisition and organization. After students have established a plan, they collect data. Students can acquire data from carefully planned experiments, accurate measurements, systematic observations, free explorations, or accidental discoveries. It doesn't matter what process produces the data; the critically important point is that students obtain data and record it. It may be necessary to reorganize and display the data for efficient analysis, often by organizing a data table. The data display is key to making sense of the science inquiry.

Making sense of data. Once students have collected and displayed their data, they need to analyze it to learn something about the natural world. In this component of the notebook, students write explanatory statements that answer the focus question. You can formalize this component by asking students to use an established protocol such as a sentence starter, or the explanation can be purely a thoughtful effort by each student. Explanations may be incorrect or incomplete at this point, but students can remedy this during the final notebook entry, when they have an opportunity to continue processing what they've learned. Unfortunately, this piece is often forgotten in the classroom during the rush to finish the lesson and move on. But without sense making and reflection (the final phase of science inquiry), students might see the lesson as a fun activity without connecting the experience to the big ideas that are being developed in the course.

Next-step strategies. The final component of an investigation brings students back to their notebooks by engaging in a next-step strategy, such as reflection and self-assessment, that moves their understanding forward. This component is the capstone on a purposeful series of experiences designed to guide students to understand the concept originally presented in the focus question. After making sense of the data, and making new claims about the topic at hand, students should go back to their earlier thinking and note their changing ideas and new findings. This reflective process helps students cement their new ideas.

A student organizes and makes sense of data in the Chemical Interactions Course.

Focusing the Investigation

Focus question. The first notebook entry in most investigations is the focus question. Focus questions are embedded in the teacher step-by-step instructions and explicitly labeled. You can write the question on the board or project it for students to transcribe into their notebooks. The focus question serves to focus students and you on the inquiry for the day. It is not always answered immediately, but rather hangs in the air while the investigation goes forward. Students always revisit their initial responses later in the investigation.

Quick write. A quick write (or quick draw) can be used in addition to a focus question. Quick writes can be completed on a quarter sheet of paper or an index card so you can collect, review, and return them to students to be taped or glued into their notebooks and used for self-assessment later in the investigation.

In the **Diversity of Life Course**, you ask,

> ➤ *What is life?*

For a quick write, students write an answer immediately, before instruction occurs. The quick write provides insight into what students think about certain phenomena before you begin instruction. When responding to the question, students should be encouraged to write down their thoughts, even if they don't feel confident in knowing the answer.

Knowing students' preconceptions will help you know what concepts need the most attention during the investigation. Make sure students date their entries for later reference. Read through students' writing and tally the important points to focus on. Quick writes should not be graded.

Planning. After students enter the focus question or complete a quick write in their notebooks, they plan their investigation. (In some investigations, planning is irrelevant to the task at hand.) Planning may be detailed or intuitive, formal or informal, depending on the requirement of the investigation. Plans might include lists (including materials, things to remember), step-by-step procedures, and experimental design. Some FOSS notebook masters guide students through a planning process specific to the task at hand.

Lists. Science notebooks often include lists of things to think about, materials to get, or words to remember. A materials list is a good organizer that helps students anticipate actions they will take. A list of variables to be controlled clarifies the purpose of an experiment. Simple lists of dates for observations or of the people responsible for completing a task may be useful.

Step-by-step procedures. Middle school students need to develop skills for writing sequential procedures. For example, in the **Chemical Interactions Course**, students write a procedure to answer these questions.

> ➤ *Is there a limit to the amount of substance that will dissolve in a certain amount of liquid?*

> ➤ *If so, is the amount that will dissolve the same for all substances?*

Students need to recall what they know about the materials, develop a procedure for accurately measuring the amount of a substance that is added to the water, and agree on a definition of "saturated." To check a procedure for errors or omissions, students can trade notebooks and attempt to follow another student's instructions to complete the task.

Experimental design. Some work with materials requires a structured experimental plan. In the **Planetary Science Course**, students pursue this focus question.

> ➤ *Are Moon craters the result of volcanoes or impacts?*

Students plan an experiment to determine what affects the size and shape of craters on the Moon. They use information they gathered during the open exploration of craters made in flour to develop a strategy for evaluating the effect of changing the variable of a projectile's height or mass. Each lab group agrees on which variable they will change and then designs a sound experimental procedure that they can refer to during the active investigation.

Data Acquisition and Organization

Because observation is the starting point for answering the focus question, data records should be

- clearly related to the focus question;

- accurate and precise;

- organized for efficient reference.

Data handling can have two subcomponents: data acquisition and data display. Data acquisition is making and recording observations (measurements). The data record can be composed of words, phrases, numbers, and drawings. Data display reorganizes the data in a logical way to facilitate thinking. The display can be a graph, chart, calendar, or other graphic organizer.

Early in a student's experience with notebooks, the record may be disorganized and incomplete, and the display may need guidance. The FOSS notebook masters are designed to help students with data collection and organization. You may initially introduce conventional data-display methods, such as those found in the FOSS notebook masters, but soon students will need opportunities to independently select appropriate data displays. As students become more familiar with collecting and organizing data, you might have them create their own records. With practice, students will become skilled at determining what form of recording to use in various situations, and how best to display the data for analysis.

Narratives. For most students, the most intuitive approach to recording data is narrative—using words, sentence fragments, and numbers in a more or less sequential manner. As students make a new observation, they record it below the previous entry, followed by the next observation, and so on. Some observations, such as a record of weather changes in the **Weather and Water Course** or the interactions of organisms in miniecosystems in the **Populations and Ecosystems Course**, are appropriately recorded in narrative form.

Drawings. A picture is worth a thousand words, and a labeled picture is even more useful. When students use a microscope to discover cells in the *Elodea* leaf and observe and draw structures of microorganisms in the **Diversity of Life Course**, a labeled illustration is the most efficient way to record data.

Charts and tables. An efficient way to record many kinds of data is a chart or table. How do you introduce this skill into the shared knowledge of the classroom? One way is to call for attention during an investigation and demonstrate how to perform the operation. Or you can let students record the data as they like, and observe their methods. There may be one or more groups that invent an appropriate table. During processing time, ask this group to share its method with the class. If no group has spontaneously produced an effective table, you might challenge the class to come up with an easier way to display the data, and turn the skill-development introduction into a problem-solving session.

With experience, students will recognize when a table or chart is appropriate for recording data. When students make similar observations on a series of objects, such as rock samples in the **Earth History Course**, a table with columns is an efficient way to organize observations for easy comparison.

Artifacts. Occasionally an investigation will produce two-dimensional artifacts that students can tape or glue directly into a science notebook. The mounted flower parts in the **Diversity of Life Course** and the sand samples card from the **Earth History Course** can become a permanent part of the record of learning.

Graphs and graphic tools. Reorganizing data into logical, easy-to-use graphic tools is typically necessary for data analysis. Graphs allow easy comparison (bar graph), quick statistical analysis of frequency data (histogram or line plot), and visual confirmation of a relationship between variables (two-coordinate graph). The **Variables and Design Course** offers many opportunities for students to collect data and organize the data into graphs. Students collect data from air trolleys traveling at different speeds, graph the data, and use the resulting graphs to understand how slope of a motion graph can indicate speed. Other graphic tools, such as Venn diagrams, pie charts, and concept maps, help students make connections.

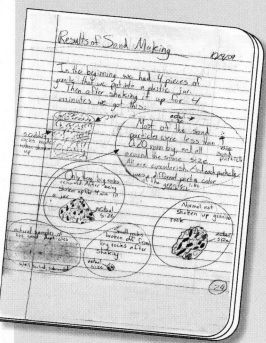

Drawing and artifact from the Earth History Course

Making Sense of Data

After collecting and organizing data, the student's next task is to answer the focus question. Students can generate an explanation as an unassisted narrative, but in many instances you might need to use supports such as the FOSS notebook masters to guide the development of a coherent and complete response to the question. Several other support structures for sense making are described below.

Development of vocabulary. Vocabulary is better introduced after students have experienced the new word(s) in context. This sequence provides a cognitive basis for students to connect accurate and precise language to their real-life experiences. Lists of new vocabulary words in the index reinforce new words and organize them for easy reference.

Data analysis. Interpreting data requires the ability to look for patterns, trends, outliers, and potential causes. Students should be encouraged to develop a habit of looking for patterns and relationships within the data collected. Frequently, this is accomplished by creating a graph with numerical data. In the **Populations and Ecosystems Course**, students review field data acquired by ecologists at Mono Lake to determine how biotic and abiotic factors affect the populations of organisms found in the lake.

Graphic organizers. Students can benefit from organizers that help them look at similarities and differences. A compare-and-contrast chart can help students make a transition from their collected data and experiences to making and writing comparisons. It is sometimes easier for students to use than a Venn diagram, and is commonly referred to as a box-and-T chart (as popularized in *Writing in Science: How to Scaffold Instruction to Support Learning*, listed in the Bibliography section).

In this strategy, students draw a box at the top of the notebook page and label it "similar" or "same." On the bottom of the notebook page, they draw a *T*. At the top of each wing of the *T*, they label the objects being compared. Students look at their data, use the *T* to identify differences for each item, and use the "similar" box to list all the characteristics that the two objects have in common. For example, a box-and-T chart comparing characteristics of extrusive and intrusive igneous rocks in the **Earth History Course** might look like this.

similar

extrusive	intrusive

Students can use the completed box-and-T chart to begin writing comparisons. It is usually easier for students to complete their chart on a separate piece of paper, so they can fill it in as they refer to their data. They affix the completed chart into their notebooks after they have made their comparisons.

Claims and evidence. A claim is an assertion about how the natural world works. Claims should always be supported by evidence—statements that are directly correlated with data. The evidence should refer to specific observations, relationships that are displayed in graphs, tables of data that show trends or patterns, dates, measurements, and so on. A claims-and-evidence construction is a sophisticated, rich display of student learning and thinking. It also shows how the data students collected is directly connected to what they learned.

Frames and prompts. One way to get students to organize their thinking is by providing sentence frames for them to complete.

- *I used to think _____ , but now I think _____ .*

- *The most important thing to remember about Moon phases is _____ .*

- *One new thing I learned about adaptation is _____ .*

Prompts also direct students to the content they should be thinking about, but provide more latitude for generating responses. For students who are learning English or who struggle with writing, assistive structures like sentence frames can help them communicate their thinking while they learn the nuances of science writing. The prompts used most often in the FOSS notebook masters take the form of questions for students to answer. In the **Weather and Water Course**, students answer the quick-write question

➤ *What causes seasons?*

After modeling an Earth/Sun system and reviewing solar angle and solar concentration, students revisit their quick write to revise and expand on their original explanations.

- *I used to think seasons were caused by _____ , but now I know _____ .*

Careful prompts scaffold students by helping them communicate their thinking but do not do the thinking for them. As students progress in communication ability, you might provide frames less frequently.

Conclusions and predictions. At the end of an investigation (major conceptual sequence), it might be appropriate for students to write a summary to succinctly communicate what they have learned. This is where students can make predictions based on their understanding of a principle or relationship. For instance, after completing the investigation of condensation and dew point in the **Weather and Water Course**, a student might predict the altitude at which clouds would form, based on weather-balloon data. Or, after examining ecosystem interactions between biotic and abiotic factors in the **Populations and Ecosystems Course**, students will predict how various human interactions could affect the ecosystem. The conclusion or prediction will frequently indicate the degree to which a student can apply new knowledge to real-world situations. A prediction can also be the springboard for further inquiry.

Generating new questions. Does the investigation connect to a student's personal interests? Does the outcome suggest a question or pique a student's curiosity? The science classroom is most exciting when students are generating their own questions for further investigation based on class or personal experiences. The notebook is an excellent place to capture students' musings and record thoughts that might otherwise be lost.

A student's revised work for the Chemical Interactions Course

Next-Step Strategies

The goal of the FOSS curriculum is for students to develop accurate, durable knowledge of the science content under investigation. Students' initial conceptions are frequently incomplete or confused, requiring additional thought to become fully functional. The science notebook is a useful place to guide reflection and revision. Typically students commit their understanding in writing and reflect in three locations.

- Explanatory narratives in notebooks
- Response sheets incorporated into the notebook
- Written work on I-Checks

These three categories of written work provide information about student learning for you *and* a record of thinking for students that they can reflect on and revise. Scientists constantly refine and clarify their ideas about how the natural world works. They read scientific articles, consult with other scientists, and attend conferences. They incorporate new information into their thinking about the subject they are researching. This reflective process can result in deeper understanding or a complete revision of thinking.

After completing one of the expositions of knowledge—a written conclusion, response sheet, or benchmark assessment—students should receive additional instruction or information via a next-step strategy. They will use this information later to complete self-assessment by reviewing their original written work, making judgments about its accuracy and completeness, and writing a revised explanation. You can use any of a number of techniques for providing the additional information to students.

- Group compare-and-share discussion
- Think/pair/share reading
- Whole-class critique of an explanation by an anonymous student
- Identifying key points for a class list
- Whole-class discussion of a presentation by one student

After one of the information-generating processes, students compare the "best answer" to their own answer and rework their explanations if they can no longer defend their original thinking. The revised statement of the science content can take one of several forms.

Students might literally revise the original writing, crossing out extraneous or incorrect bits, inserting new or improved information, and completing the passage. At other times, students might reflect on their original work and, after drawing and dating a line of learning (see below), might redraft their explanation from scratch, producing their best explanation of the concept.

During these self-assessment processes, students have to think actively about every aspect of their understanding of the concept and organize their thoughts into a coherent, logical narrative. The learning that takes place during this process is powerful. The relationships between the several elements of the concept become unified and clarified.

The notebook is the best tool for students when preparing for benchmark assessment, such as an I-Check or posttest. Students don't necessarily have the study skills needed to prepare on their own, but using teacher-guided tasks such as key points and traffic lights will turn the preparation process into a valuable exercise. These same strategies can be used after a benchmark assessment when you identify further areas of confusion or misconceptions you want to address with students. Here are four helpful next-step, or self-assessment, strategies.

Line of learning. One technique many teachers find useful in the reflective process is the line of learning. After students have conducted an investigation and entered their initial explanations, they draw and date a line under their original work. As students share ideas and refine their thinking during class discussion, additional experimentation, reading, and teacher feedback, encourage them to make new entries under the line of learning, adding to or revising their original thinking. If the concept is elusive or complex, a second line of learning, followed by more processing and revising, may be appropriate.

The line of learning is a reminder to students that learning is an ongoing process with imperfect products. It points out places in that process where a student made a stride toward full understanding. And the psychological security provided by the line of learning reminds students that they can always draw another line of learning and revise their thinking again. The ability to look back in the science notebook and see concrete evidence of learning gives students confidence and helps them become critical observers of their own learning.

A line of learning used with the Planetary Science Course

Traffic lights. In the traffic-lights strategy, students use color to self-assess and indicate how well they understand a concept that they are learning. Green means that the student feels that he or she has a good understanding of the concept. Yellow means that the student is still a bit unsure about his or her understanding. Red means that the student needs help; he or she has little or no understanding of the concept. Students can use colored pencils, markers, colored dots, or colored cards to indicate their understanding. They can mark their own work and then indicate their level of understanding by a show of hands or by holding up colored cards. This strategy gives students practice in self-assessment and helps you monitor students' current understanding. You should follow up by looking at student work to ensure that they actually do understand the content that they marked with green.

Three C's. Another approach to revision is to apply the three C's—confirm, correct, complete—to the original work. Students indicate ideas that were correct with a number or a color, code statements needing correction with a second number or color, and assign a third number or color to give additional information that completes the entry.

Key points. Students do not necessarily connect the investigative experience with the key concepts and processes taught in the lesson. It is essential to give students an opportunity to reflect on their experiences and find meaning in those experiences. They should be challenged to use their experiences and data to either confirm or reject their current understanding of the natural world. As students form supportable ideas about a concept, those ideas should be noted as key points, posted in the room, and written in their notebooks. New evidence, to support or clarify an idea, can be added to the chart as the course progresses. If an idea doesn't hold up under further investigation, a line can be drawn through the key point to indicate a change in thinking. A key-points activity is embedded near the end of each investigation to help students organize their thinking and prepare for benchmark assessment.

USING NOTEBOOKS TO IMPROVE STUDENT LEARNING

Notebook entries should not be graded. Research has shown that more learning occurs when students get only comments on written work in their notebooks, not grades or a combination of comments and grades.

If your school district requires a certain number of grades each week, select certain work products that you want to grade and have students turn in that work separate from the notebook. After grading, return the piece to students to insert into their notebooks, so that all their work is in one place.

It may be difficult to stop using grades or a rubric for notebook assessment. But providing feedback that moves learning forward, however difficult, has benefits that make it worth the effort. The key to using written feedback for formative assessment is to make feedback timely and specific, and to provide time for students to act on the feedback by revising or correcting work right in their own notebook.

Teacher Feedback

Student written work often exposes weaknesses in understanding—or so it appears. It is important for you to find out if the flaw results from poor understanding of the science or from imprecise communication. You can use the notebook to provide two types of feedback to the student: to ask for clarification or additional information, and to ask probing questions that will help students move forward in their thinking. Respecting the student's space is important, so rather than writing directly in the notebook, attach a self-stick note, which can be removed after the student has taken appropriate action.

The most effective forms of feedback relate to the content of the work. Here are some examples.

> ➤ *You wrote that seasons are caused by Earth's tilt. Does Earth's tilt change during its orbit?*

> ➤ *What evidence can you use to support your claim that Moon craters are caused by impacts? Hint: Think of our experiments in class.*

Nonspecific feedback, such as stars, pluses, smiley faces, and "good job!", or ambiguous critiques, such as "try again," "put more thought into this," and "not enough," are less effective and should not be used. Feedback that guides students to think about the content of their work and gives suggestions for how to improve are productive instructional strategies.

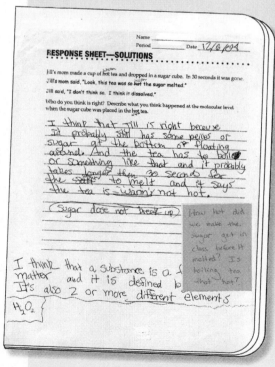

Feedback given during the Chemical Interactions Course

Here are some appropriate generic feedback questions to write or use verbally while you circulate in the class.

➤ *What vocabulary have you learned that will help you describe* _____ ?

➤ *Can you include an example from class to support your ideas?*

➤ *Include more detail about* _____ .

➤ *Check your data to make sure this is accurate.*

➤ *What do you mean by* _____ ?

➤ *When you record your data, what unit should you use?*

When students return to their notebooks and respond to the feedback, you will have additional information to help you discriminate between learning and communication difficulties. Another critical component of teacher feedback is providing comments to students in a timely manner, so that they can review their work before engaging in benchmark assessment or moving on to other big ideas in the course.

In middle school, you face the challenge of having a large number of students. This may mean collecting a portion of students' notebooks on alternate days. Set a specific focus for your feedback, such as a data table or conclusion, so you aren't trying to look at everything every time.

To help students improve their writing, you might have individuals share notebook entries aloud in their collaborative groups, followed by feedback from a partner or the group. This valuable tool must be very structured to create a safe environment, including ground rules about acceptable feedback and comments.

A good way to develop these skills is to model constructive feedback with the class, using a student-work sample from a notebook. Use a sample from a previous year with the name and any identifying characteristics removed. Project it for the class to practice giving feedback.

Formative Assessment

With students recording more of their thinking in an organized notebook, you have a tool to better understand the progress of students and any misconceptions that are typically not revealed until the benchmark assessment. One way to monitor student progress is during class while they are responding to a prompt. Circulate from group to group, and read notebook entries over students' shoulders. This is a good time to have short conversations with individuals or small groups to gain information about the level of student understanding. Take care to respect the privacy of students who are not comfortable sharing their work during the writing process.

If you want to look at work that is already completed in the notebook, ask students to open their notebooks to the page that you want to review and put them in a designated location. Or consider having students complete the work on a separate piece of paper or an index card. Students can leave a blank page in their notebooks, or label it with a header as a placeholder, until they get the work back and tape it or glue it in place. This makes looking at student work much easier, and the record of learning that the student is creating in the notebook remains intact.

When time is limited, you might select a sample of students from each class, alternating the sample group each time, to get a representative sample of student thinking. This is particularly useful following a quick write.

Once you have some information about student thinking, you can make teaching decisions about moving ahead to a benchmark assessment, going back to a previous concept, or spending more time making sense of a concept. Benchmark assessments can also be used as formative assessment. You might choose to administer an I-Check, score the assessment to find problem areas, and then revisit critical concepts before moving on to the next investigation. Students can use reflection and self-assessment techniques to revisit and build on their original exam responses.

Quick writes for the Waves Course written on index cards

DERIVATIVE PRODUCTS

On occasion, you might ask students to produce science projects of various kinds: summary reports, detailed explanations, end-of-course projects, oral reports, or posters. Students should use their notebooks as a reference when developing their reports. You could ask them to make a checklist of science concepts and pieces of evidence, with specific page references, extracted from their notebooks. They can then use this checklist to ensure that all important points have been included in the derivative work.

The process of developing a project has feedback benefits, too. While students are developing projects using their notebooks, they have the opportunity to self-monitor the organization and content of the notebook. This offers valuable feedback on locating and extracting useful information. You might want to discuss possible changes students would make next time they start a new science notebook.

Homework is another form of derivative product, as it is an extension of the experimentation started in class. Carefully selected homework assignments enhance students' science learning. Homework suggestions and/or extension activities are included at the end of each investigation. For example, in the **Heredity and Adaptation Course**, after using an online activity in class to predict genetic variation, students are asked to complete a follow-up online simulation as homework. In the **Electromagnetic Force Course**, after students test properties of magnets in class, they are asked to look for examples of magnets in household objects outside of the classroom.

Science-Centered Language Development in Middle School

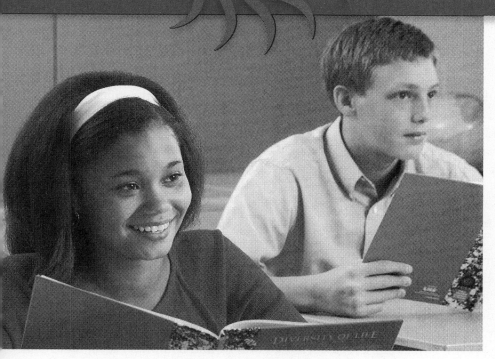

Reading and writing are inextricably linked to the very nature and fabric of science, and, by extension, to learning science.

Stephen P. Norris and Linda M. Phillips, "How Literacy in Its Fundamental Sense Is Central to Scientific Literacy"

Contents

INTRODUCTION

In this chapter, we explore the ways reading, writing, speaking, and listening are interwoven in effective science instruction at the secondary level. To engage fully in the enterprise of science and engineering, students must record and communicate observations and explanations, and read about and discuss the discoveries and ideas of others. This becomes increasingly challenging at the secondary level. Texts become more complex; writing requires fluency of academic language, including domain-specific vocabulary. Here we identify strategies that support sense making. The active investigations, student science notebooks, *FOSS Science Resources* readings, multimedia, and formative assessments provide rich contexts in which students develop and exercise thinking processes and communication skills. Students develop scientific literacy through experiences with the natural world around them in real and authentic ways and use language to inquire, process information, and communicate their thinking about the objects, organisms, and phenomena they are studying. We refer to the acquisition and building of language skills necessary for scientific literacy as science-centered language development.

Language plays two crucial roles in science learning: (1) it facilitates the communication of conceptual and procedural knowledge, questions, and propositions (external; public), and (2) it mediates thinking, a process necessary for understanding (internal; private). These are also the ways scientists use language: to communicate with one another about their inquiries, procedures, and understandings; to transform their observations into ideas; and to create meaning and new ideas from their work and the work of others. For students, language development is intimately involved in their learning about the natural world. Active-learning science provides a real and engaging context for developing literacy; language-arts skills and strategies support conceptual development and scientific and engineering practices. For example, the skills and strategies used for reading comprehension, writing expository text, and oral discourse are applied when students are recording their observations, making sense of science content, and communicating their ideas. Students' use of language improves when they discuss, write, and read about the concepts explored in each investigation.

We begin our exploration of science and language by focusing on language functions and how specific language functions are used in science to facilitate information acquisition and processing (thinking). Then we address issues related to the specific language domains—speaking and listening, writing, and reading. Each section addresses

- how skills in that domain are developed and exercised in FOSS science investigations;

- literacy strategies that are integrated purposefully into the FOSS investigations; and

- suggestions for additional literacy strategies that both enhance student learning in science and develop or exercise academic-language skills.

Following the domain discussions is a section on science-vocabulary development, with scaffolding strategies for supporting all learners. The last section covers language-development strategies specifically for English learners.

▶ **NOTE**
The term *English learners* refers to students who are learning to understand English. This includes students who speak English as a second language and native English speakers who need additional support to use language effectively.

THE ROLE OF LANGUAGE IN SCIENTIFIC AND ENGINEERING PRACTICES

Language functions are the purpose for which speech or writing is used and involve both vocabulary and grammatical structure. Understanding and using language functions appropriately is important in effective communication. Students use numerous language functions in all disciplines to mediate communication and facilitate thinking (e.g., they plan, compare, discuss, apply, design, draw, and provide evidence).

In science, language functions facilitate scientific and engineering practices. For example, when students are *collecting data*, they are using language functions to identify, label, enumerate, compare, estimate, and measure. When students are *constructing explanations*, they are using language functions to analyze, communicate, discuss, evaluate, and justify.

A Framework for K–12 Science Education (National Research Council 2012) states that "Students cannot comprehend scientific practices, nor fully appreciate the nature of scientific knowledge itself, without directly experiencing the practices for themselves." Each of these scientific and engineering practices uses multiple language functions. Often, these language functions are part of an internal dialogue weighing the merits of various explanations—what we call thinking. The more language functions with which we are facile, the more effective and creative our thinking can be.

The scientific and engineering practices are listed below, along with a sample of the language functions that are exercised when students are effectively engaged in that practice. (Practices are bold; language functions are italic.)

Asking questions and defining problems

- *Ask* questions about objects, organisms, systems, and events in the natural and human-made world (science).

- *Ask* questions to *define* and *clarify* a problem, *determine criteria* for solutions, and *identify* constraints (engineering).

Planning and carrying out investigations

- *Plan* and conduct investigations in the laboratory and in the field to gather appropriate data (*describe* procedures, *determine* observations to *record*, *decide* which variables to control) or to gather data essential for *specifying* and *testing* engineering designs.

Examples of Language Functions

Analyze
Apply
Ask
Clarify
Classify
Communicate
Compare
Conclude
Construct
Critique
Describe
Design
Develop
Discuss
Distinguish
Draw
Enumerate
Estimate
Evaluate
Experiment
Explain
Formulate
Generalize
Group
Identify
Infer
Interpret
Justify
Label
List
Make a claim
Measure
Model
Observe
Organize
Plan
Predict
Provide evidence
Reason
Record
Represent
Revise
Sequence
Solve
Sort
Strategize
Summarize
Synthesize

Analyzing and interpreting data

- Use a range of tools (numbers, words, tables, graphs, images, diagrams, equations) to *organize* observations (data) in order to *identify* significant features and patterns.

Developing and using models

- Use models to help *develop explanations, make predictions*, and *analyze* existing systems, and *recognize* strengths and limitations of the models.

Using mathematics and computational thinking

- Use mathematics and computation to *represent* physical variables and their relationships.

Constructing explanations and designing solutions

- *Construct* logical explanations of phenomena, or *propose solutions* that incorporate current understanding or a model that represents it and is consistent with the available evidence.

Engaging in argument from evidence

- *Defend* explanations, *formulate evidence* based on data, *examine* one's own understanding in light of evidence offered by others, and collaborate with peers in searching for explanations.

Obtaining, evaluating, and communicating information

- *Communicate* ideas and the results of inquiry—orally and in writing—with tables, diagrams, graphs, and equations and in *discussion* with peers.

Research supports the claim that when students are intentionally using language functions in thinking about and communicating in science, they improve not only science content knowledge, but also language-arts and mathematics skills (Ostlund, 1998; Lieberman and Hoody, 1998). Language functions play a central role in science as a key cognitive tool for developing higher-order thinking and problem-solving abilities that, in turn, support academic literacy in all subject areas.

Here is an example of how an experienced teacher can provide an opportunity for students to exercise language functions in FOSS. In the **Planetary Science Course**, one piece of content we expect students to have acquired by the end of the course is

- The lower the angle at which light strikes a surface, the lower the density of the light energy.

The scientific practices the teacher wants the class to focus on are *developing and using models* and *constructing explanations*.

The language functions students will exercise while engaging in these scientific practices include *comparing, modeling, analyzing*, and *explaining*. The teacher understands that these language functions are appropriate to the purpose of the science investigation and support the *Common Core State Standards for English Language Arts and Literacy in Science* (CCSS), in which grades 6–8 students will "write arguments focused on discipline-specific content . . . support claim[s] with logical reasoning and relevant, accurate data and evidence that demonstrate an understanding of the topic" (National Governors Association Center for Best Practices, Council of Chief State School Officers, 2010).

- Students will *compare* the area covered by the same beam of light (from a flashlight) at different angles to *explain* the relationship between the angle and density of light energy.

The teacher can support the use of language functions by providing structures such as sentence frames.

- As _____, then _____.

  As the angle increases, then the light beam becomes smaller and more circular.

- When I changed _____, then _____.

  When I changed the angle of the light beam, then the concentration of light hitting the floor changed.

- The greater/smaller _____, the _____.

  The smaller the angle of the light beam, the more the light beam spread out.

- I think _____, because _____.

  I think the smaller spot of light receives more energy than the larger spot because the light concentration is greatest when light shines directly down on a surface and there is no beam spreading.

> ▶ **CCSS NOTE**
> This example supports
> CCSS.ELA-Literacy.WHST.6–8.1.b.

SPEAKING AND LISTENING DOMAIN

The FOSS investigations are designed to engage students in productive oral discourse. Talking requires students to process and organize what they are learning. Listening to and evaluating peers' ideas calls on students to apply their knowledge and to sharpen their reasoning skills. Guiding students in instructive discussions is critical to the development of conceptual understanding of the science content and the ability to think and reason scientifically. It also addresses a key middle school CCSS Speaking and Listening anchor standard that students "engage effectively in a range of collaborative discussions (one-on-one, in groups, and teacher-led) with diverse partners on [grade-level] topics, texts, and issues, building on others' ideas and expressing their own clearly."

▶ **CCSS NOTE**
This example supports
CCSS.ELA-Literacy.SL.6.1,
CCSS.ELA-Literacy.SL.7.1, and
CCSS.ELA-Literacy.SL.8.1.

FOSS investigations start with a discussion—a review to activate prior knowledge, presentation of a focus question, or a challenge to motivate and engage active thinking. During the active investigation, students talk with one another in small groups, share their observations and discoveries, point out connections, ask questions, and start to build explanations. The discussion icon in the sidebar of the *Investigations Guide* indicates when small-group discussions should take place.

Throughout the activity, the *Investigations Guide* indicates where it is appropriate to pause for whole-class discussions to guide conceptual understanding. The *Investigations Guide* provides you with discussion questions to help stimulate student thinking and support sense making. At times, it may be beneficial to use sentence frames or standard prompts to scaffold the use of effective language functions and structures. Allowing students a few minutes to write in their notebooks prior to sharing their answers also helps those who need more time to process and organize their thoughts.

On the following pages are some suggestions for providing structure to those discussions and for scaffolding productive discourse when needed. Using the protocols that follow will ensure inclusion of all students in discussions.

▶ **NOTE**
Additional notebook strategies can be found in the Science Notebooks in Middle School chapter in *Teacher Resources* and online at www.FOSSweb.com.

Partner and Small-Group Discussion Protocols

Whenever possible, give students time to talk with a partner or in a small group before conducting a whole-class discussion. This provides all students with a chance to formulate their thinking, express their ideas, practice using the appropriate science vocabulary, and receive input from peers. Listening to others communicate different ways of thinking about the same information from a variety of perspectives helps students negotiate the difficult path of sense making for themselves.

Dyads. Students pair up and take turns either answering a question or expressing an idea. Each student has 1 minute to talk while the other student listens. While student A is talking, student B practices attentive listening. Student B makes eye contact with student A, but cannot respond verbally. After 1 minute, the roles reverse.

Here's an example from the **Chemical Interactions Course**. After reviewing the results of eight reactions recorded in their notebooks, you ask students to pair up and take turns sharing which two substances they think constitute the mystery mixture and their reasons for selecting those two. The language objective is for students to compare their test results and make inferences based on their observations and what they know about chemical reactions (orally and in writing). These sentence frames can be written on the board to scaffold student thinking and conversation.

- I think the two substances in the mystery mixture are _____ and _____.

- My evidence is _____.

Partner parade. Students form two lines facing each other. Present a question, an idea, an object, or an image as a prompt for students to discuss. Give them 1 minute to greet the person in front of them and discuss the prompt. After 1 minute, call time. Have the first student in one of the lines move to the end of the line, and have the rest of the students in that line shift one step sideways so that everyone has a new partner. (Students in the other line do not move.) Give students a new prompt to discuss for 1 minute with their new partners. This can also be done by having students form two concentric circles. After each prompt, the inner circle rotates.

For example, when students are just beginning the **Earth History Course** investigation on igneous rock, you may want to assess prior knowledge about Earth's layers. Give each student a picture from an assortment of related images such as volcanoes, magma, a diagram of Earth's layers, crystals, and so forth, and have students line up facing

Partner and Small-Group Discussion Protocols
- *Dyads*
- *Partner parade*
- *Put in your two cents*

each other in two lines or in concentric circles. For the first round, ask, "What do you observe in the image on your card?" For the second round, ask, "What can you infer from the image?" For the third round, ask, "What questions do you have about the image?" The language objective is for students to describe their observations, infer how the landform formed, and reflect upon and relate any experiences they may have had with a similar landform. These sentence frames can be used to scaffold student discussion.

- I observe _____, _____, and _____.
- I think this shows _____ because _____.
- I wonder _____.

Put in your two cents. For small-group discussions, give each student two pennies or similar objects to use as talking tokens. Each student takes a turn putting a penny in the center of the table and sharing his or her idea. Once all have shared, each student takes a turn putting in the other penny and responding to what others in the group have said. For example,

- I agree (or don't agree) with _____ because _____.

Here's an example from the **Diversity of Life Course**. In their notebooks, students have recorded the amount of water lost from their vials containing celery with and without leaves. They discover a discrepancy in the amount of water lost and the mass of the celery. Where did the water go? Students are struggling to form an explanation. The language objective is for students to compare their results and infer that there is a relationship between the amount of water lost and the number of leaves the celery has. You give each student two pennies, and in groups of four, they take turns putting in their two cents. For the first round, each student answers the question "Where did the water go?" They use the frame

- I think the water _____.
- My evidence is _____.

On the second round, each student states whether he or she agrees or disagrees with someone else in the group and why, using the sentence frame.

Whole-Class Discussion Supports

The whole-class discussion is a critical part of sense making. After students have had the active learning experience and have talked with their peers in pairs and/or small groups, sharing their observations with the whole class sets the stage for developing conventional explanatory models. Discrepant events, differing results, and other surprises are discussed, analyzed, and resolved. It is important that students realize that science is a process of finding out about the world around them. This is done through asking questions, testing ideas, forming explanations, and subjecting those explanations to logical scrutiny, that is, argumentation. Leading students through productive discussion helps them connect their observations and the abstract symbols (words) that represent and explain those observations. Whole-class discussion also provides an opportunity for you to interject an accurate and precise verbal summary as a model of the kind of thinking you are seeking. Facilitating effective whole-class discussions takes skill, practice, a shared set of norms, and patience. In the long run, students will have a better grasp of the content and will improve their ability to think independently and communicate effectively.

Norms should be established so that students know what is expected during science discussions.

- Science content and practices are the focus.

- Everyone participates (speaking and listening).

- Ideas and experiences are shared, accepted, and valued. Everyone is respectful of one another.

- Claims are supported by evidence.

- Challenges (debate and argument) are part of the quest for complete understanding.

A variety of whole-class discussion techniques can be introduced and practiced during science instruction that address the CCSS Speaking and Listening standards for students to "present claims and findings [e.g., argument, narrative, informative, summary presentations], emphasizing salient points in a focused, coherent manner with relevant evidence, sound valid reasoning, and well-chosen details; use appropriate eye contact, adequate volume, and clear pronunciation."

For example, during science talk, students are reminded to practice attentive listening, stay focused on the speaker, ask questions, and respond appropriately. In addition, in order for students to develop and practice their reasoning skills, they need to know the language forms

Whole-Class Discussion Supports
- *Sentence frames*
- *Guiding questions*

TEACHING NOTE

Let students know that scientists change their minds based on new evidence. It is expected that students will revise their thinking, based on evidence presented in discussions.

▶ **CCSS NOTE**
This example supports CCSS.ELA-Literacy.SL.6.4, CCSS.ELA-Literacy.SL.7.4, and CCSS.ELA-Literacy.SL.8.4, (grade 8 quoted here).

---

and structures and the behaviors used in evidence–based debate and argument, such as using data to support claims, disagreeing respectfully, and asking probing questions (Winokur and Worth, 2006).

Explicitly model the language structures appropriate for active discussions, and encourage students to use them when responding to guiding questions and during science talks.

Sentence frames. These samples can be posted as a scaffold as students develop their reasoning and oral participation skills.

- I think _____, because _____.
- I predict _____, because _____.
- I claim _____; my evidence is _____.
- I agree with _____ that _____.
- My idea is similar/related to _____'s idea.
- I learned/discovered/heard that _____.
- <Name> explained _____ to me.
- <Name> shared _____ with me.
- We decided/agreed that _____.
- Our group sees it differently, because _____.
- We have different observations/results. Some of us found that _____. One group member thinks that _____.
- We had a different approach/idea/solution/answer: _____.

Guiding questions. The Investigations Guide provides questions to help concentrate student thinking on the concepts introduced in the investigation. Guiding questions should be used during the whole-class discussion to facilitate sense making. Here are some other open-ended questions that help guide student thinking and promote discussion.

- What did you notice when _____?
- What do you think will happen if _____?
- How might you explain _____? What is your evidence?
- What connections can you make between _____ and _____?

Whole-Class Discussion Protocols

The following tried-and-true participation protocols can be used to enhance whole-class discussions. The purpose of these protocols is to increase meaningful participation by giving all students access to the discussion, allowing students time to think (process), and providing a context for motivation and engagement.

Think-pair-share. When asking for a response to a question posed to the class, give students a minute to think silently. Then, have students pair up with a partner to exchange thoughts before you call on a student to share his or her ideas with the whole class.

Pick a stick. Write each student's name on a craft stick, and keep the sticks handy in a cup at the front of the room. When asking for responses, randomly pick a stick, and call on that student to start the discussion. Continue to select sticks as you continue the discussion. Your name can also be on a stick in the cup. To keep students on their toes, put the selected sticks into a smaller cup hidden inside the larger cup out of view of students. That way students think they may be called again.

Whip around. Each student takes a quick turn sharing a thought or reaction. Questions are phrased to elicit quick responses that can be expressed in one to five words (e.g., "Give an example of a stored-energy source." "What does the word *heat* make you think of?").

Commit and toss. Have students write a response to a question or prompt on a loose piece of paper (Keeley, 2008). Next, tell everyone to crumple up the paper into a ball and toss it to another student. Continue tossing for a few minutes, and then call for students to stop, grab a ball, and read the response silently. Responses can then be shared with partners, small groups, or the whole class. This activity allows students to answer anonymously, so they may be willing to share their thinking more openly.

Group posters. Have small groups design and graphically record their investigation data and conclusions on a quickly generated poster to share with the whole class.

Whole-Class Discussion Protocols
- *Think-pair-share*
- *Pick a stick*
- *Whip around*
- *Commit and toss*
- *Group posters*

Cup within a cup pick-a-stick container

WRITING DOMAIN

Information processing is enhanced when students engage in informal writing. When allowed to write expressively without fear of being scorned for incorrect spelling or grammar, students are more apt to organize and express their thoughts in different ways that support their own sense making. Writing in science promotes use of science and engineering practices, thereby developing a deeper engagement with the science content. This type of informal writing also provides a springboard for more formal derivative science writing (Keys, 1999).

Science Notebooks

The science notebook is an effective tool for enhancing learning in science and exercising various forms of writing. Notebooks provide opportunities both for expressive writing (students craft explanatory narratives that make sense of their science experiences) and for practicing informal technical writing (students use organizational structures and writing conventions). Students learn to communicate their thinking in an organized fashion while engaging in the cognitive processes required to develop concepts and build explanations. Having this developmental record of learning also provides an authentic means for assessing students' progress in both scientific thinking and communication skills.

Developing Writing for Literacy in Science

Using student science notebooks in science instruction provides opportunities to address the CCSS for Writing in Science. Grades 6–8 students "write routinely over extended time frames (time for research, reflection, and revision) and shorter time frames (a single sitting or a day or two) for a range of tasks, purposes, and audiences." In addition to providing a structure for recording and analyzing data, notebooks serve as a reference tool from which students can draw information in order to produce derivative products, that is, more formal science writing pieces that have a specific purpose and format. CCSS focus on three text types that students should be writing in science: argument, informational/explanatory writing, and narrative writing. These text types are used in science notebooks and can be developed into derivative products such as reports, articles, brochures, poster boards, electronic presentations, letters, and so forth. Following is a description of these three text types and examples that may be used with FOSS investigations to help students build scientific literacy.

▶ **NOTE**
For more information about supporting science-notebook development, see the Science Notebooks in Middle School chapter.

▶ **CCSS NOTE**
This example supports CCSS.ELA-Literacy.W.10.

Engaging in Argument

In science, middle school students make claims in the form of statements or conclusions that answer questions or address problems. CCSS Appendix A describes that for students to use "data in a scientifically acceptable form, students marshal evidence and draw on their understanding of scientific concepts to argue in support of their claims." Applying the literacy skills necessary for this type of writing concurrently supports the development of critical science and engineering practices—most notably, engaging in argument. According to *A Framework for K–12 Science Education*, upon which the Next Generation Science Standards (NGSS) are based, middle school students are expected to construct a convincing argument that supports or refutes claims for explanations about the natural and designed world in these ways.

In FOSS, this type of writing makes students' thinking visible. Both informally in their notebooks and formally on assessments, students use deductive and inductive reasoning to construct and defend their explanations. In this way, students deepen their science understanding and exercise the language functions necessary for higher-level thinking, for example, comparing, synthesizing, evaluating, and justifying. To support students in both oral and written argumentation, use the questions and prompts in the *Investigations Guide* that encourage students to use evidence, models, and theories to support their arguments. In addition, be prepared for those teachable moments that provide the perfect stage for spontaneous scientific debate. Here are some general questions to help students deepen their writing.

- Why do you agree or disagree with _____?

- How would you prove/disprove _____?

- What data did you use to make that conclusion that _____?

- Why was it better that _____?

Here are the ways engaging in written argument are developed in the FOSS investigations and can be extended through formal writing.

Response sheets. The FOSS response sheets give students practice in constructing arguments by providing hypothetical situations where they have to apply what they have learned in order to evaluate a claim. For example, one of the response sheets in the **Planetary Science Course** asks students to respond to three students' explanations for the seasons. Students write a paragraph to each student with the

Engaging in Argument
- *Response sheets*
- *Think questions*
- *I-Checks and surveys/posttests*
- *Persuasive writing*

▶ **CCSS NOTE**
This example supports
CCSS.ELA-Literacy.W.1.

purpose of changing his or her thinking. In order to refute each claim, students must evaluate the validity of the statements and construct arguments based on evidence from the data they've collected during the investigations and logical reasoning that supports their explanation for what causes seasons.

Think questions. Interactive reading in *FOSS Science Resources* is another opportunity for students to engage in written argumentation. Articles include questions that support reading comprehension and extend student thinking about the science content. Asking students to make a claim and provide evidence to support it encourages the use of language functions necessary for higher-level thinking such as evaluating, applying, and justifying. For example, in *FOSS Science Resources: Planetary Science*, students are asked to respond to the following question: Why do you think there are so few craters on Earth and so many on the Moon? After discussion with their peers, students can hone their argumentation skills by writing an argument that answers the question and is supported by the evidence in the *FOSS Science Resources* book as well as data recorded from their experience making model craters.

I-Checks and surveys/posttests. Like the FOSS response sheets, some test items assess students' ability to make a claim and provide evidence to support it. One way is to provide students with data and have them make a claim based on that data and evidence from their prior investigations. Their argument should use logical reasoning to support their ideas. For example, in **Planetary Science**, students are shown images taken from two different planets. They are told that one has a thick atmosphere and the other has no atmosphere. They are asked which image they think came from a planet with an atmosphere and why. Using the images, they can see evidence of craters, and they can draw on their own experiences as well as knowledge acquired through other sources to piece together a logical argument.

Persuasive writing. Formal writing gives students the opportunity to summarize, explain, apply, and evaluate what they have learned in science. It also provides a purpose and audience that motivate students to produce higher-level writing products. The objective of persuasive writing is to convince the reader that a stated interpretation of data is worthwhile and meaningful. In addition to supporting claims with evidence and using logical argument, the writer also uses persuasive techniques such as a call to action. Students can use their informal notebook entries to form the basis of formal persuasive writing in a variety of formats, such as essays, letters, editorials, advertisements, award nominations, informational pamphlets, and petitions. Animal habitats, energy use, weather patterns, landforms, and water sources are just a few science topics that can generate questions and issues for persuasive writing.

Here is a sample of writing frames that can be used to introduce and scaffold persuasive writing (modified from Gibbons, 2002).

Title: _____

The topic of this discussion is _____.

My opinion (position, conclusion) is _____.

There are <number> reasons why I believe this to be true.

First, _____.

Second, _____.

Finally, _____.

On the other hand, some people think _____.

I have also heard people say _____.

However, my claim is that _____ because _____.

Informational/Explanatory Writing

- *Writing frames*
- *Recursive cycle*

▶ **CCSS NOTE**

Designing, recording, and following procedures in FOSS courses supports CCSS.ELA-Literacy.RST.6–8.3.

▶ **CCSS NOTE**

This example supports CCSS.ELA-Literacy.W.2.

Informational/Explanatory Writing

Informational and explanatory writing requires students to examine and convey complex ideas and information clearly and accurately through the effective selection, organization, and analysis of content. In middle school science, this includes writing scientific procedures and experiments. Described in CCSS Appendix A, informational/ explanatory writing answers the questions, What type? What are the components? What are the properties, functions, and behaviors? How does it work? What is happening? Why? In FOSS, this type of writing takes place informally in science notebooks, where students are recording their questions, plans, procedures, data, and answers to the focus questions. It also supports sense making as students attempt to convey what they know in response to questions and prompts, using language functions such as identifying, comparing and contrasting, explaining cause-and-effect relationships, and sequencing.

As an extension of the notebook entries, students can apply their content knowledge to publish formal products such as letters, definitions, procedures, newspaper and magazine articles, posters, pamphlets, and research reports. Strategies such as the writing process (plan, draft, edit, revise, and share) and writing frames (modeling and guiding the use of topic sentences, transition and sequencing words, examples, explanations, and conclusions) can be used to scaffold and help students develop proficiency in science writing.

Writing frames. Here are samples of writing frames (modified from Wellington and Osborne, 2001).

Description

Title: _____

(Identify) The part of the _____ I am describing is the _____.

(Describe) It consists of _____.

(Explain) The function of these parts is _____.

(Example) This drawing shows _____.

Explanation

Title: _____

I want to explain why (how) _____.

An important reason for why (how) this happens is that _____.

Another reason is that _____.

I know this because _____.

Recursive cycle. An effective method for extending students' science learning through writing is the recursive cycle of research (Bereiter, 2002). This strategy emphasizes writing as a process for learning, similar to the way students learn during the active science investigations.

1. Decide on a problem or question to write about.

2. Formulate an idea or a conjecture about the problem or question.

3. Identify a remedy or an answer, and develop a coherent discussion.

4. Gather information (from an experiment, science notebooks, *FOSS Science Resources*, FOSSweb multimedia, books, Internet, interviews, videos, etc.).

5. Reevaluate the problem or question based on what has been learned.

6. Revise the idea or conjecture.

7. Make presentations (reports, posters, electronic presentations, etc.).

8. Identify new needs, and make new plans.

This process can continue for as long as new ideas and questions occur, or students can present a final product in any of the suggested formats.

Narrative Writing

Narrative writing conveys an experience to the reader, usually with sensory detail and a sequence of events. In middle school science, students learn the importance of writing narrative descriptions of their procedures with enough detail and precision to allow others to replicate the experiment. Science also provides a broad landscape of stimulating material for stories, songs, biographies, autobiographies, poems, and plays. Students can enrich their science learning by using organisms or objects as characters; describing habitats and environments as settings; and writing scripts portraying various systems, such as weather patterns, states of matter, and the water, rock, or life cycle.

▶ **CCSS NOTE**
This example supports CCSS.ELA-Literacy.W.7.

▶ **NOTE**
Human characteristics should not be given to organisms (anthropomorphism) in science investigations, only in literacy extensions.

▶ **CCSS NOTE**
This example supports CCSS.ELA-Literacy.W.3.

READING DOMAIN

Reading is an integral part of science learning. Just as scientists spend a significant amount of their time reading each other's published works, students need to learn to read scientific text—to read effectively for understanding, with a critical focus on the ideas being presented.

The articles in *FOSS Science Resources* facilitate sense making as students make connections to the science concepts introduced and explored during the active investigations. Concept development is most effective when students are allowed to experience organisms, objects, and phenomena firsthand before engaging the concepts in text. The text and illustrations help students make connections between what they have experienced concretely and the abstract ideas that explain their observations.

FOSS Science Resources provides students with clear and coherent explanations, ways of visualizing important information, and different perspectives to examine and question. As students apply these strategies, they are, in effect, using some of the same scientific thinking processes that promote critical thinking and problem solving. In addition, the text provides a level of complexity appropriate for middle schoolers to develop high-level reading comprehension skills. This development requires support and guidance as students grapple with more complex dimensions of language meaning, structure, and conventions. To become proficient readers of scientific and other academic texts, students must be armed with an array of reading comprehension strategies and have ample opportunities to practice and extend their learning by reading texts that offer new language, new knowledge, and new modes of thought.

Oral discourse and writing are critical for reading comprehension and for helping students make sense of the active investigations. Use the suggested prompts, questions, and strategies in the *Investigations Guide* to support comprehension as students read from *FOSS Science Resources*. For most of the investigation parts, the articles are designed to follow the active investigation and are interspersed throughout the course. This allows students to acquire the necessary background knowledge in context through active experience before tackling the wider-ranging content and relationships presented in the text. Breakpoints in the readings are suggested in the *Investigations Guide* to support student conceptual development. Some questions make connections between the reading and the student's class experience. Other questions help the students consider the writer's intent. Additional strategies for reading are derived from the seven essential strategies that readers use to help them understand what they read (Keene and Zimmermann, 2007).

▶ **CCSS NOTE**

The use of *FOSS Science Resources* supports CCSS.ELA-Literacy.RST.6–8.10.

▶ **CCSS NOTE**

Reading breakpoints in the *Investigations Guide* support CCSS.ELA-Literacy.RST.6–8.8.

- Monitor for meaning: discover when you know and when you don't know.

- Use and create schemata: make connections between the novel and the known; activate and apply background knowledge.

- Ask questions: generate questions before, during, and after reading that reach for deeper engagement with the text.

- Determine importance: decide what matters most, what is worth remembering.

- Infer: combine background knowledge with information from the text to predict, conclude, make judgments, and interpret.

- Use sensory and emotional images: create mental images to deepen and stretch meaning.

- Synthesize: create an evolution of meaning by combining understanding with knowledge from other texts/sources.

Reading Comprehension Strategies

Below are some strategies that enhance the reading of expository texts in general and have proven to be particularly helpful in science. Read and analyze the articles beforehand in order to guide students through the text structures and content more effectively.

Build on background knowledge. Activating prior knowledge is critical for helping students make connections between what they already know and new information. Reading comprehension improves when students have the opportunity to think, discuss, and write about what they know about a topic before reading. Review what students learned from the active investigation, provide prompts for making connections, and ask questions to help students recall past experiences and previous exposure to concepts related to the reading.

Create an anticipation guide. Create true-or-false statements related to the key ideas in the reading selection. Ask students to indicate if they agree or disagree with each statement before reading, then have them read the text, looking for the information that supports their true-or-false claims. Anticipation guides connect students to prior knowledge, engage them with the topic, and encourage them to explore their own thinking. To provide a challenge for advanced students, have them come up with the statements for the class.

Draw attention to vocabulary. Check the article for bold faced words students may not know. Review the science words that are already defined in students' notebooks. For new science and nonscience

Reading Comprehension Strategies
- *Build on background knowledge*
- *Create an anticipation guide*
- *Draw attention to vocabulary*
- *Preview the text*
- *Turn and talk*
- *Jigsaw text reading*
- *Note making*
- *Summarize and synthesize*
- *3-2-1*
- *Write reflections*
- *Preview and predict*
- *SQ3R*

▶ **CCSS NOTE**

The example of reviewing what students learned from the active investigation supports CCSS.ELA-Literacy.RST.6–8.9.

▶ **CCSS NOTE**

This example supports CCSS.ELA-Literacy.RST.6–8.4.

vocabulary words that appear in the reading, have students predict their meanings before reading. During the reading, have students use strategies such as context clues and word structure to see if their predictions were correct. This strategy activates prior knowledge and engages students by encouraging analytical participation with the text.

Preview the text. Give students time to skim through the selection, noting subheads, before reading thoroughly. Point out the particular structure of the text and what discourse markers to look for. For example, most *FOSS Science Resources* articles are written as cause and effect, problem and solution, question and answer, comparison and contrast, description, or sequence. Students will have an easier time making sense of the text if they know what text structure to look for. Model and have students practice analyzing these different types of expository text structures by looking for examples, patterns, and discourse markers. For example, let's look at a passage from *FOSS Science Resources: Planetary Science*.

> An eclipse of the Moon occurs when Earth passes exactly between the Moon and the Sun. [cause and effect] The Moon moves into Earth's shadow during a lunar eclipse. At the time of a full lunar eclipse, Earth's shadow completely covers the disk of the Moon. [description] This is how Earth, the Moon, and the Sun are aligned for a lunar eclipse to be observed. [photograph] Why don't we see a lunar eclipse every month? [question and answer] Because of the tilt of the Moon's orbit around the Earth, Earth's shadow does not fall on the Moon in most months.

Point out how the text in *FOSS Science Resources* is organized (titles, headings, subheadings, questions, and summaries) and if necessary, review how to use the table of contents, glossary, and index. Explain how to scan for formatting features that provide key information (such as boldface type and italics, captions, and framed text) and graphic features (such as tables, graphs, photographs, maps, diagrams, and charts) that help clarify, elaborate, and explain important information in the reading.

While students preview the article, have them focus on the questions that appear in the text, as well as questions at the end of the article. Encourage students to write down questions they have that they think the article will answer.

Turn and talk. When reading as a whole class, stop at key points and have students share their thinking about the selection with the student sitting next to them or with their collaborative group. This strategy helps students process the information and allows everyone to participate in the discussion. When reading in pairs, encourage

▶ **NOTE**
Discourse markers are words or phrases that relate one idea to another. Examples are *however, on the other hand,* and *second.*

▶ **CCSS NOTE**
This example supports CCSS.ELA-Literacy.RST.6-8.5 and CCSS.ELA-Literacy.RST.6-8.6.

▶ **CCSS NOTE**
This example supports CCSS.ELA-Literacy.RST.6-8.7.

students to stop and discuss with their partners. One way to encourage engagement and understanding during paired reading is to have students take turns reading aloud a paragraph or section on a certain topic. The one who is listening then summarizes the meaning conveyed in the passage.

Jigsaw text reading. Students work together in small groups (expert teams) to develop a collective understanding of a text. Each expert team is responsible for one portion of the assigned text. The teams read and discuss their portions to gain a solid understanding of the key concepts. They might use graphic organizers to refine and organize the information. Each expert team then presents its piece to the rest of the class. Or form new jigsaw groups that consist of at least one representative from each expert team. Each student shares with the jigsaw group what their team learned from their particular portion of the text. Together, the participants in the jigsaw group fit their individual pieces together to create a complete picture of the content in the article.

Note making. The more students interact with a reading, the better their understanding. Encourage students to become active readers by asking them to make notes as they read. Studies have shown that note making—especially paraphrasing and summarizing—is one of the most effective means for understanding text (Graham and Herbert, 2010; Applebee, 1984). Some investigation parts include notebook sheets that match pages in *FOSS Science Resources*. This allows students to highlight and underline important points, add notes in the margins, and circle words they do not know. Students can also annotate the article by writing thoughts and questions on self-stick notes. Using symbols or codes can help facilitate comprehension monitoring. Here are some possible symbols students can use to communicate their thinking as they interact with text. (Harvey, 1998).

★	interesting
BK	background knowledge
?	question
C	confusing
I	important
L	learning something new
W	wondering
S	surprising

▶ **CCSS NOTE**
This example supports
CCSS.ELA-Literacy.RST.6–8.10.

▶ **CCSS NOTE**
This example supports
CCSS.ELA-Literacy.RST.6–8.1 and
CCSS.ELA-Literacy.RST.6–8.2.

Students can also use a different set of symbols while making notes about connections: the readings in *FOSS Science Resources* incorporate the active learning that students gain from the investigations, so that they can make authentic text-to-self (T-S) connections. In other words, what they read reminds them of firsthand experiences, making the article more engaging and easier to understand. Text-to-text (T-T) connections are notes students make when they discover a new idea that reminds them of something they've read previously in another text. Text-to-world (T-W) connections involve the text and more global everyday connections to students' lives.

You can model note-making strategies by displaying a selection of text, using a projection system, a document camera, or an interactive whiteboard. As you read the text aloud, model how to write comments on self-stick notes, and use a graphic organizer in a notebook to enhance understanding.

FACTS	QUESTIONS
L – When a planet crosses in front of a star, it causes a dimming of the star called a "blink."	Can we observe any other transits in our lifetime?
I – Exoplanets can be detected by observing star blinks.	How many more exoplanets are there?
C – Earth's transit would cause a very small dimming, but it would be long in brightness.	What does a "long drop in brightness" mean?

An example of annotated text from *FOSS Science Resources: Planetary Science*

Graphic organizers help students focus on extracting the important information from the reading and analyzing relationships between concepts. This can be done by simply having students make columns in their notebooks to record information and their thinking (Harvey and Goudvis, 2007). Here are two examples.

Notes	Thinking

Facts	Questions	Responses

▶ CCSS NOTE
This example supports CCSS.ELA-Literacy.RST.6–8.1 and CCSS.ELA-Literacy.RST.6–8.2.

Summarize and synthesize. Model how to pick out the important parts of the reading selection. Paraphrasing is one way to summarize. Have students write summaries of the reading, using their own words. To scaffold the learning, use graphic organizers to compare and contrast, group, sequence, and show cause and effect. Another method is to have students make two columns in their notebooks. In one column, they record what is important, and in the other, they record their personal responses (what the reading makes them think about). When writing summaries, tell students,

- *Pick out the important ideas.*

- *Restate the main ideas in your own words.*

- *Keep it brief.*

3-2-1. This graphic-organizer strategy gives students the opportunity to synthesize information and formulate questions they still have regarding the concepts covered in an article. In their notebooks, students write three new things they learned, two interesting things worth remembering and sharing, and one question that occurred to them while reading the article. Other options might include three facts, two interesting ideas, and one insight about themselves as learners; three key words, two new ideas, and one thing to think about (modified from Black Hills Special Services Cooperative, 2006).

Write reflections. After reading, ask students to review their notes in their notebooks to make any additions, revisions, or corrections to what they recorded during the reading. This review can be facilitated by using a line of learning. Students draw a line under their original conclusion or under their answer to a question posed at the end of an article. They add any new information as a new narrative entry. The line of learning indicates that what follows represents a change of thinking.

Preview and predict. Instruct students to independently preview the article, directing attention to the illustrations, photos, boldfaced words, captions, and anything else that draws their attention. Working with a partner, students discuss and write three things they think they will learn from the article. Have partners verbally share their list with another pair of students. The group of four can collaborate to generate one list. Groups report their ideas, and together you create a class list on chart paper.

Read the article aloud, or have students read with a partner aloud or silently. Referring to the preview/prediction list, discuss what students learned. Have them record the most important thing they learned from the reading for comparison with the predictions.

SQ3R. Survey, Question, Read, Recall, Reflect strategy provides an overall structure for before, during, and after reading. Students begin by surveying or previewing the text, looking for features that will help them make predictions about the content. Based on their surveys, students develop questions to answer as they read. They read the selections looking for answers to their questions. Next, they recall what they have learned by retelling a partner and/or recording what they've learned. Finally, they reflect on what they have learned, check to see that they've answered their questions sufficiently, and add any new ideas. Below is a chart students can use to record the SQ3R process in their notebooks.

S Survey	Q Question	R Read	R Recall	R Reflect
Scan the text and record important information.	Ask questions about the subject and what you already know.	Record answers to your questions after you read.	Retell what you learned in your own words.	Did you answer your questions? Record new ideas and comments.

Struggling Readers

For students reading below grade level, the strategies listed on the previous pages can be modified to support reading comprehension by integrating scaffolding strategies such as read-alouds and guided reading. Breaking the reading down into smaller chunks, providing graphic organizers, and modeling reading comprehension strategies can also help students who may be struggling with the text. For additional strategies for English learners, see the supported-reading strategy in the English-Language Development section of this chapter.

Interactive reading aloud. Reading aloud is an effective strategy for enhancing text comprehension. It offers opportunities to model specific reading comprehension strategies and allows students to concentrate on making sense of the content. When modeling, share the thinking processes used to understand the reading (questioning, visualizing, comparing, inferring, summarizing, etc.), then have students share what they observed you thinking about as an active reader.

Guided reading. While the rest of the class is reading independently or in small groups, pull a group aside for a guided reading session. Before reading, review vocabulary words from the investigation and ask questions to activate prior knowledge. Have students preview the text to make predictions, ask questions, and think about text structure. Review reading comprehension strategies they will need to use (monitoring for understanding, asking questions, summarizing, synthesizing, etc.). As students read independently, provide support where needed. Ask questions and provide prompts to guide comprehension. (See the list below for additional strategies.) After reading, have students reflect on what strategies they used to help them understand the text and make connections to the investigation.

- While reading, look for answers to questions and confirm predictions.
- Study graphics, such as pictures, graphs, and tables.
- Reread captions associated with pictures, graphs, and tables.
- Note all italicized and boldfaced words or phrases.
- Reduce reading speed for difficult passages.
- Stop and reread parts that are not clear.
- Read only a section at a time, and summarize after each section.

Struggling Readers
- *Interactive reading aloud*
- *Guided reading*

SCIENCE-VOCABULARY DEVELOPMENT

Words play two critically important functions in science. First and most important, we play with ideas in our minds, using words. We present ourselves with propositions—possibilities, questions, potential relationships, implications for action, and so on. The process of sorting out these thoughts involves a lot of internal conversation, internal argument, weighing options, and complex linguistic decisions. Once our minds are made up, communicating that decision, conclusion, or explanation in writing or through verbal discourse requires the same command of the vocabulary. Words represent intelligence; acquiring the precise vocabulary and the associated meanings is key to successful scientific thinking and communication.

The words introduced in FOSS investigations represent or relate to fundamental science concepts and should be taught in the context of the investigation. Many of the terms are abstract and are critical to developing science content knowledge and scientific and engineering practices. The goal is for students to use science vocabulary in ways that demonstrate understanding of the concepts the words represent—not to merely recite scripted definitions. The most effective strategies for science-vocabulary development help students make connections to what they already know. These strategies focus on giving new words conceptual meaning through experience; distinguishing between informal, everyday language and academic language; and using the words in meaningful contexts.

Building Conceptual Meaning through Experience

In most instances, students should be presented with new words when they need to know them in the context of the active experience. Words such as *kinetic energy, atmospheric pressure, chemical reaction, photosynthesis*, and *transpiration* are abstract and conceptually loaded. Students will have a much better chance of understanding, assimilating, and remembering the new word (or new meaning) if they can connect it with a concrete experience.

The vocabulary icon appears in the sidebar when students are prompted to record new words in their notebook. The words that appear in bold are critical to understanding the concepts or scientific practices students are learning and applying in the investigation.

When you introduce a new word, students should

- Hear it: students listen as you model the correct contextual use and pronunciation of the word;

- See it: students see the new word written out;

- Say it: students use the new word when discussing their observations and inferences; and

- Write it: students use the new words in context when they write in their notebooks.

Bridging Informal Language to Science Vocabulary

FOSS investigations are designed to tap into students' inquisitive nature and their excitement of discovery in order to encourage lively discussions as they explore materials in creative ways. There should be a lot of talking during the investigations! Your role is to help students connect informal language to the vocabulary used to express specific science concepts. As you circulate during active investigations, you continually model the use of science vocabulary. For example, as students are examining a leaf under the microscope, they will say, "I can see little mouths." You might respond, "Yes, those mouthlike openings are called stomates. They are pores that open and close." Below are some strategies for validating students' conversational language while developing their familiarity with and appreciation for science vocabulary.

Cognitive-content dictionaries. Choose a term that is critical for conceptual understanding of the science investigation. Have students write the term, predict its meaning, write the final meaning after class discussion (using primary language or an illustration), and use the term in a sentence.

Cognitive-Content Dictionary	
New term	kinetic energy
Prediction (clues)	something that moves a lot
Final meaning	motion energy
How I would use it in a sentence	Fast-moving particles have more kinetic energy than slow-moving particles.

Bridging Informal Language to Science Vocabulary
- *Cognitive-content dictionaries*
- *Concept maps*
- *Semantic webs*
- *Word associations*
- *Word sorts*

Concept maps. Select six to ten related science words. Have students write them on self-stick notes or cards. Have small groups discuss how the words are related. Students organize words in groups and glue them down or copy them on a sheet of paper. Students draw lines between the related words. On the lines, they write words describing or explaining how the concept words are related.

Semantic webs. Select a vocabulary word, and write it in the center of a piece of paper (or on the board for the whole class). Brainstorm a list of words or ideas that are related to the first word. Group the words and concepts into several categories, and attach them to the central word with lines, forming a web (modified from Hamilton, 2002).

Word associations. In this brainstorming activity, you say a word, and students respond by writing the first word that comes to mind. Then students share their words with the class. This activity builds connections to students' prior frames of reference.

Word sorts. Have students work with a partner to make a set of word cards using new words from the investigation. Have them group the words in different ways, for example, synonyms, root words, and conceptual connections.

Using Science Vocabulary in Context

For a new vocabulary word to become part of a student's functional vocabulary, he or she must have ample opportunities to hear and use it. Vocabulary terms are used in the activities through teacher talk, whole-class and small-group discussions, writing in science notebooks, readings, and assessments. Other methods can also be used to reinforce important vocabulary words and phrases.

Word wall. Use chart paper to record science content and procedural words as they come up during and after the investigations. Students will use this word wall as a reference.

Drawings and diagrams. For English learners and visual learners, use a diagram to review and explain abstract content. Ahead of time, draw an illustration lightly, almost invisibly, with pencil on chart paper. You can do this easily by projecting the image onto the paper. When it's time for the investigation, trace the illustration with markers as you introduce the words and phrases to students. Students will be amazed by your artistic ability.

Science Vocabulary Strategies

- *Word wall*
- *Drawings and diagrams*
- *Cloze activity*
- *Word wizard*
- *Word analysis/word parts*
- *Breaking apart words*
- *Possible sentences*
- *Reading*
- *Glossary*
- *Index*
- *Poems, chants, and songs*

Cloze activity. Structure sentences for students to complete, leaving out the vocabulary words. This can be done as a warm-up with the words from the previous day's lesson. Here's an example from the **Earth History Course**.

> Teacher: *The removal and transportation of loose earth materials is called* _____.
>
> Students: *Erosion.*

Word wizard. Tell students that you are going to lead a word activity. You will be thinking of a science vocabulary word. The goal is to figure out the word. Provide hints that have to do with parts of a definition, root word, prefix, suffix, and other relevant components. Students work in teams of two to four. Provide one hint, and give teams 1 minute to discuss. One team member writes the word on a piece of paper or on the whiteboard, using dark marking pens. Each team holds up its word for only you to see. After the third clue, reveal the word, and move on to the next word. Here's an example.

> 1. *Part of the word means green.*
> 2. *They are found in plant cells.*
> 3. *They look like tiny green spheres or ovals.*
>
> *The word is* **chloroplasts.**

Word analysis/word parts. Learning clusters of words that share a common origin can help students understand content-area texts and connect new words to familiar ones. Here's an example: *geology, geologist, geological, geography, geometry, geophysical.* This type of contextualized teaching meets the immediate need of understanding an unknown word while building generative knowledge that supports students in figuring out difficult words for future reading.

Breaking apart words. Have teams of two to four students break a word into prefix, root word, and suffix. Give each team different words, and have each team share the parsed elements of the word with the whole class. Here's an example.

> *photosynthesis*
>
> Prefix = *photo*: meaning light
>
> Root = *synthesis*: meaning to put together

Possible sentences. Here is a simple strategy for teaching word meanings and generating class discussion.

1. Choose six to eight key concept words from the text of an article in *FOSS Science Resources*.

2. Choose four to six additional words that students are more likely to know something about.

3. Put the list of ten to fourteen words on the board or project it. Provide brief definitions as needed.

4. Ask students to devise sentences that include two or more words from the list.

5. On chart paper, write all sentences that students generate, both coherent and otherwise.

6. Have students read the article from which the words were extracted.

7. Revisit students' sentences, and discuss whether the sentences are sensible based on the passage or how they could be modified to be more coherent.

Reading. After the active investigation, students continue to develop their understanding of the vocabulary words and the concepts those words represent by listening to you read aloud, reading with a partner, or reading independently. Use strategies discussed in the Reading Domain section to encourage students to articulate their thoughts and practice the new vocabulary.

Glossary. Emphasize the vocabulary words students should be using when they answer the focus question in their science notebooks. The glossary in *FOSS Science Resources* or on FOSSweb can be used as a reference.

Index. Have students create an index at the back of their notebooks. There they can record new vocabulary words and the notebook page where they defined and used the new words for the first time in the context of the investigation.

Poems, chants, and songs. As extensions or homework assignments, ask students to create poems, raps, chants, or songs, using vocabulary words from the investigation.

▶ **NOTE**
See the Science Notebooks in Middle School chapter for an example of an index.

ENGLISH-LANGUAGE DEVELOPMENT

Active investigations, together with ample opportunities to develop and use language, provide an optimal learning environment for English learners. This section highlights opportunities for English-language development (ELD) in FOSS investigations and suggests other best practices for facilitating both the learning of new science concepts and the development of academic literacy. For example, the hands-on structure of FOSS investigations is essential for the conceptual development of science content knowledge and the habits of mind that guide and define scientific and engineering practices. Students are engaged in concrete experiences that are meaningful and that provide a shared context for developing understanding—critical components for language acquisition.

When getting ready for an investigation, review the steps and determine the points where English learners may require scaffolds and where the whole class might benefit from additional language-development supports. One way to plan for ELD integration in science is to keep in mind four key areas: prior knowledge, comprehensible input, academic language development, and oral practice. The ELD chart lists examples of universal strategies for each of these components that work particularly well in teaching science.

▶ NOTE
English-language development refers to the advancement of students' ability to read, write, and speak English.

English-Language Development (ELD)	
Activating prior knowledge	**Using comprehensible input**
• Inquiry chart	• Content objectives
• Circle map	• Multiple exposures
• Observation poster	• Visual aids
• Quick write	• Supported reading
• Kit inventory	• Procedural vocabulary
Developing academic language	**Providing oral practice**
• Language objectives	• Small-group discussions
• Sentence frames	• Science talk
• Word wall, word cards, drawings	• Oral presentations
• Concept maps	• Poems, chants, and songs
• Cognitive content dictionaries	• Teacher feedback

> **NOTE**
Language forms and structures are the internal grammatical structure of words and how those words go together to make sentences.

Students acquiring English benefit from scaffolds that support the language forms and functions necessary for the academic demands of the science course, that is, accessing science text, participating in productive oral discourse, and engaging in science writing. The table at the end of this section (starting on page 38) provides a resource to help students organize their thinking and structure their speaking and writing in the context of the science and engineering practices. The table identifies key language functions exercised during FOSS investigations and provides examples of sentence frames students can use as scaffolds.

For example, if students are planning an investigation to learn more about insect structures and behaviors, the language objective might be "Students plan and design an investigation that answers a question about the hissing cockroach's behavior." For students who need support, a sentence frame that prompts them to identify the variables in the investigation would provide language forms and structures appropriate for planning their investigation. As a scaffold, sentence frames can also help them write detailed narratives of their procedure. Here's an example from the table.

> **NOTE**
The complete table appears at the end of the English-Language Development section starting on page 38.

Language functions	Language objectives	Sentence frames
Planning and carrying out investigations		
Design Sequence Strategize Evaluate	Plan controlled experiments with multiple trials. Identify independent variable and dependent variable. Discuss, describe, and evaluate the methods for collecting data.	To find out _____, I will change _____. I will not change _____. I will measure _____. I will observe _____. I will record the data by _____. First, I will _____, and then I will _____. To learn more about _____, I will need _____ to _____.

Activating Prior Knowledge

When an investigation engages a new concept, students first recall and discuss familiar situations, objects, or experiences that relate to and establish a foundation for building new knowledge and conceptual understanding. Eliciting prior knowledge also supports learning by motivating interest, acknowledging culture and values, and checking for misconceptions and prerequisite knowledge. This is usually done in the first steps of Guiding the Investigation in the form of a discussion, presentation of new materials, or a written response to a prompt. The tools outlined below can also be used before beginning an investigation to establish a familiar context for launching into new material.

Circle maps. Draw two concentric circles on chart paper. In the smaller circle, write the topic to be explored. In the larger circle, record what students already know about the subject. Ask students to think about how they know or learned what they already know about the topic. Record the responses outside the circles. Students can also do this independently in their science notebooks.

An example of a circle map

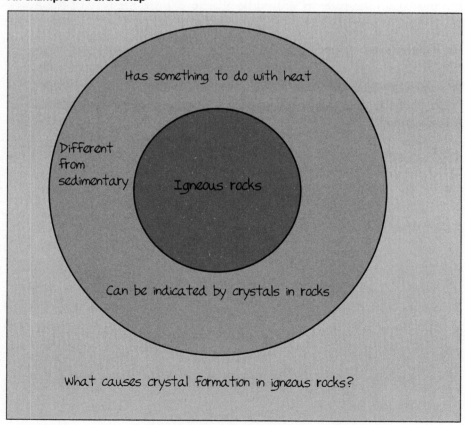

Activating Prior Knowledge
- *Circle maps*
- *Observation posters*
- *Quick writes*
- *Kit inventories*

Observation posters. Make observation posters by gluing or taping pictures and artifacts relevant to the module or a particular investigation onto pieces of blank chart paper or poster paper. Hang them on the wall in the classroom, and have students rotate in small groups to each poster. At each poster, students discuss their observations with their partners or small groups and then record (write or draw) an observation, a question, a prediction, or an inference about the pictures as a contribution to the commentary on the poster.

As a variation on this strategy, give a set of pictures to each group to pass around. Have them choose one and write what they notice, what they infer, and questions they have in their notebooks.

Quick writes. Ask students what they know about the topic of the investigation. Responses can be recorded independently as a quick write in notebooks and then shared collaboratively. Do not correct misconceptions initially. Periodically revisit the quick-write ideas as a whole class, or have students review their notebook entries to correct, confirm, or complete their original thoughts as they acquire new information (possibly using a line of learning). At the conclusion of the investigation, students should be able to express mastery of the new conceptual material.

Kit inventories. Introduce each item from the FOSS kit used in the investigation, and ask students questions to get them thinking about what each item is and where they may have seen it before. Have them describe the objects and predict how they will be used in the investigation.

Comprehending Input

To initiate their own sense making, students must be able to access the information presented to them. We refer to this ability as comprehending input. Students must understand the essence of new ideas and concepts before beginning to construct new scientific meaning. The strategies for comprehensible input used in FOSS ensure that the instruction is understandable while providing students with the opportunity to grapple with new ideas and the critically important relationships between concepts. Additional tools such as repetition, visual aids, emphasis on procedural vocabulary, and auditory reinforcement can also be used to enhance comprehensible input for English learners.

Content objectives. The focus question for each investigation part frames the activity objectives—what students should know or be able to do at the end of the part. Making the learning objectives clear and explicit prepares English learners to process the delivery of new information, and helps you maintain the focus of the investigation. Write the focus question on the board, have students read it aloud and transcribe it into their science notebooks, and have students answer the focus question at the end of the investigation part. You then check their responses for understanding.

Multiple exposures. Repeat an activity in an analogous but slightly different context, ideally one that incorporates elements that are culturally relevant to students. For example, as a homework assignment for landforms, have students interview their parents about landforms common in the area of their ancestry.

Visual aids. On the board or chart paper, write out the steps for conducting the investigation. This provides a visual reference. Include illustrations if necessary. Use graphic representations (illustrations drawn and labeled in front of students) to review the concepts explored in the active investigations. In addition to the concrete objects in the kit, use realia to augment the activity, to help English learners build understanding and make cultural connections. Graphic organizers (webs, Venn diagrams, T-tables, flowcharts, etc.) aid comprehension by helping students see how concepts are related.

Supported reading. In addition to the reading comprehension strategies suggested in the Reading Domain section of this chapter, English learners can also benefit from methods such as front-loading key words, phrases, and complex text structures before reading; using

Comprehending Input
- *Content objectives*
- *Multiple exposures*
- *Visual aids*
- *Supported reading*
- *Procedural vocabulary*

Procedural Vocabulary

Add
Analyze
Assemble
Attach
Calculate
Change
Classify
Collect
Communicate
Compare
Connect
Construct
Contrast
Demonstrate
Describe
Determine
Draw
Evaluate
Examine
Explain
Explore
Fill
Graph
Identify
Illustrate
Immerse
Investigate
Label
List
Measure
Mix
Observe
Open
Order
Organize
Pour
Predict
Prepare
Record
Represent
Scratch
Separate
Sort
Stir
Subtract
Summarize
Test
Weigh

preview-review (main ideas are previewed in the primary language, read in English, and reviewed in the primary language); and having students use sentence frames specifically tailored to record key information and/or graphic organizers that make the content and the relationship between concepts visually explicit from the text as they read.

Procedural vocabulary. Make sure students understand the meaning of the words used in the directions for an investigation. These may or may not be science-specific words. Use techniques such as modeling, demonstrating, and body language (gestures) to explain procedural meaning in the context of the investigation. The words students will encounter in FOSS include those listed in the sidebar. To build academic literacy, English learners need to learn the multiple meanings of these words and their specific meanings in the context of science.

Developing Academic Language

As students learn the nuances of the English language, it is critical that they build proficiency in academic language in order to participate fully in the cognitive demands of school. *Academic language* refers to the more abstract, complex, and specific aspects of language, such as the words, grammatical structure, and discourse markers that are needed for higher cognitive learning. FOSS investigations introduce and provide opportunities for students to practice using the academic vocabulary needed to access and engage with science ideas.

Language objectives. Consider the language needs of English learners and incorporate specific language-development objectives that will support learning the science content of the investigation, such as a specific way to expand use of vocabulary by looking at root words, prefixes, and suffixes; a linguistic pattern or structure for oral discussion and writing; or a reading comprehension strategy. Recording in science notebooks is a productive way to optimize science learning and language objectives. For example, in the **Earth History Course**, one language objective might be "Students will apply techniques for rock observations to compare and contrast sedimentary and igneous rocks. They will discuss and record their observations in their notebooks in an organized manner."

Vocabulary development. The Science-Vocabulary Development section in this chapter describes the ways science vocabulary is introduced and developed in the context of an active investigation and suggests methods and strategies that can be used to support vocabulary development during science instruction. In addition to science vocabulary, students need to learn the nonscience vocabulary that facilitates deeper understanding and communication skills. Words such as *release*, *convert*, *beneficial*, *produce*, *receive*, *source*, and *reflect* are used in the investigations and *FOSS Science Resources* and are frequently used in other content areas. Learning these academic-vocabulary words gives students a more precise and complex way of practicing and communicating productive thinking. Consider using the strategies described in the Science-Vocabulary Development section to explicitly teach targeted, high-leverage words that can be used in multiple ways and that can help students make connections to other words and concepts. Sentence frames, word walls, concept maps, and cognitive-content dictionaries are strategies that have been found to be effective with academic-vocabulary development.

Scaffolds That Support Science and Engineering Practices

Language functions	Language objectives	Sentence frames
Asking questions and defining problems		
Inquire Define a problem	Ask questions to solicit information about phenomena, models, or unexpected results; determine the constraints and criteria of a problem.	I wonder why ____ . What happens when ____? What if ____? What does ____? What can ____? What would happen if ____? How does ____ affect ____? How can I find out if ____? Which ____ is better for ____?
Planning and carrying out investigations		
Design Sequence Strategize Evaluate	Plan controlled experiments with multiple trials. Identify independent variable and dependent variable. Discuss, describe, and evaluate the methods for collecting data.	To find out ____, I will change ____. I will not change ____. I will measure ____. I will observe ____. I will record the data by ____. First, I will ____, and then I will ____. To learn more about ____, I will need ____ to ____.

Language functions	Language objectives	Sentence frames
Planning and carrying out investigations *(continued)*		
Describe	Write narratives using details to record sensory observations and connections to prior knowledge.	I observed/noticed _____. When I touch the _____, I feel _____. It smells _____. It sounds _____. It reminds me of _____, because _____.
Organize Compare Classify	Make charts and tables: use a T-table or chart for recording and displaying data.	The table compares _____ and _____
Sequence Compare	Record changes over time, and describe cause-and-effect relationships.	At first, _____, but now _____. We saw that first _____, then _____, and finally _____. When I _____, it _____. After I _____, it _____.
Draw Label Identify	Draw accurate and detailed representations; identify and label parts of a system using science vocabulary, with attention to form, location, color, size, and scale.	The diagram shows _____. _____ is shown here. _____ is _____ times bigger than _____. _____ is _____ times smaller than _____ .
Analyzing and interpreting data		
Enumerate Compare Represent	Use measures of variability to analyze and characterize data; decide when and how to use bar graphs, line plots, and two-coordinate graphs to organize data.	The mean is _____. The median is _____. The mode is _____. The range is _____. The x-axis represents _____ and the y-axis represents _____. The units are expressed in _____.

Language functions	Language objectives	Sentence frames
Analyzing and interpreting data *(continued)*		
Compare Classify Sequence	Use graphic organizers and narratives to express similarities and differences, to assign an object or action to the category or type to which it belongs, and to show sequencing and order.	This _____ is similar to _____ because _____. This _____ is different from _____ because _____. All these are _____ because _____. _____, _____, and _____ all have/are _____.
Analyze	Use graphic organizers, narratives, or concept maps to identify part/whole or cause–and–effect relationships. Express data in qualitative terms such as more/fewer, higher/lower, nearer/farther, longer/shorter, and increase/decrease; and quantitatively in actual numbers or percentages.	The _____ consists of _____. The _____ contains _____. As _____, then _____. When I changed _____, then _____ happened. The more/less _____, then _____.
Developing and using models		
Represent Predict Explain	Construct and revise models to predict, represent, and explain.	If _____, then _____, therefore _____. The _____ represents _____ . _____ shows how _____. You can explain _____ by _____.

Language functions	Language objectives	Sentence frames
Using mathematics and computational thinking		
Symbolize Measure Enumerate Estimate	Use mathematical concepts to analyze data.	The ratio of _____ is _____ to _____. The average is _____. Looking at _____, I think there are _____. My prediction is _____.
Constructing explanations and designing solutions		
Infer Explain	Construct explanations based on evidence from investigations, knowledge, and models; use reasoning to show why the data are adequate for the explanation or conclusion.	I claim that _____. I know this because _____. Based on _____, I think _____. As a result of _____, I think _____. The data show _____, therefore, _____. I think _____ means _____ because _____. I think _____ happened because _____.
Provide evidence	Use qualitative and quantitative data from the investigation as evidence to support claims. Use quantitative expressions using standard metric units of measurement such as cm, mL, °C.	My data show _____. My evidence is _____. The relationship between the variables is _____. The model of _____ shows that _____.

Language functions	Language objectives	Sentence frames
Engaging in argument from evidence		
Discuss Persuade Synthesize Negotiate Suggest	Use oral and written arguments supported by evidence and reasoning to support or refute an argument for a phenomenon or a solution to a problem.	I think ___ because___. I agree/disagree with ___ because_____. What you are saying is _____. What do you think about _____? What if _____? I think you should try ___. Another way to interpret the data is _____.
Critique Evaluate Reflect	Evaluate competing design solutions based on criteria; compare two arguments from evidence to identify which is better.	____ makes more sense because ____. ____ is a better design _____ because it ____. Comparing ___ to ___ shows that _____. One discrepancy is ____. ____ is inconsistent with _____. Another way to determine _____ is to _____. I used to think ____, but now I think ____. I have changed my thinking about ____. I am confused about ____ because ____. I wonder ____.
Obtaining, evaluating, and communicating information		
(This practice includes all functions described in the other practices above.)		

REFERENCES

Applebee, A. 1984. "Writing and Reasoning." *Review of Educational Research* 54 (Winter): 577–596.

Bereiter, C. 2002. *Education and Mind in the Knowledge Age.* Hillsdale, NJ: Erlbaum.

Black Hills Special Services Cooperative. 2006. "3-2-1 Strategy." In *On Target: More Strategies to Guide Learning.* Rapid City, SD: South Dakota Department of Education.

Gibbons, P. 2002. *Scaffolding Language, Scaffolding Learning.* Portsmouth, NH: Heinemann.

Graham, S., and M. Herbert. 2010. *Writing to Read: Evidence for How Writing Can Improve Reading.* New York: Carnegie.

Hamilton, G. 2002. *Content-Area Reading Strategies: Science.* Portland, ME: Walch Publishing.

Harvey, S. 1998. *Nonfiction Matters: Reading, Writing, and Research in Grades 3–8.* Portland, ME: Stenhouse.

Harvey, S., and A. Goudvis. 2007. *Strategies That Work: Teaching Comprehension for Understanding and Engagement.* Portland, ME: Stenhouse.

Keeley, P. 2008. *Science Formative Assessment: 75 Practical Strategies for Linking Assessment, Instruction, and Learning.* Thousand Oaks, CA: Corwin Press.

Keene, E., and S. Zimmermann. 2007. *Mosaic of Thought: The Power of Comprehension Strategies.* 2nd ed. Portsmouth, NH: Heinemann.

Keys, C. 1999. *Revitalizing Instruction in Scientific Genres: Connecting Knowledge Production with Writing to Learn in Science.* Athens: University of Georgia.

Lieberman, G. A., and L. L. Hoody. 1998. *Closing the Achievement Gap: Using the Environment as an Integrating Context for Learning.* San Diego, CA: State Education and Environment Roundtable.

National Governors Association Center for Best Practices, Council of Chief State School Officers. 2010. *Common Core State Standards for English Language Arts & Literacy in History/Social Studies, Science, and Technical Subjects.* Washington, DC: National Governors Association Center for Best Practices, Council of Chief State School Officers.

National Research Council. 2012. *A Framework for K–12 Science Education: Practices, Crosscutting Concepts, and Core Ideas.* Committee on a Conceptual Framework for New K–12 Science Education Standards. Board on Science Education, Division of Behavioral and Social Sciences and Education. Washington, DC: The National Academies Press.

▶ **NOTE**
For additional resources and updated references, go to FOSSweb.

Norris, S. P., and L. M. Phillips. 2003. "How Literacy in Its Fundamental Sense Is Central to Scientific Literacy." *Science Education* 87 (2).

Ostlund, K. 1998. "What the Research Says about Science Process Skills: How Can Teaching Science Process Skills Improve Student Performance in Reading, Language Arts, and Mathematics?" *Electronic Journal of Science Education* 2 (4).

Wellington, J., and J. Osborne. 2001. *Language and Literacy in Science Education.* Buckingham, UK: Open University Press.

Winokur, J., and K. Worth. 2006. "Talk in the Science Classroom: Looking at What Students and Teachers Need to Know and Be Able to Do." In *Linking Science and Literacy in the K–8 Classroom*, ed. R. Douglas, K. Worth, and W. Binder. Arlington, VA: NSTA Press.

FOSSweb and Technology

Adding Magnetic Fields

Start

Contents

INTRODUCTION

FOSSweb technology is an integral part of the **Electromagnetic Force Course**. It provides students with the opportunity to access and interact with simulations, images, video, and text—digital resources that can enhance their understanding of life science concepts. Different sections of digital resources are incorporated into each investigation during the course. Each use is marked with the technology icon in the *Investigations Guide*. You will sometimes use the digital resources to make presentations to the class. At other times, individuals or small groups of students will work with the digital resources to review concepts or reinforce their understanding.

The FOSSweb components are not optional. To prepare to use these digital resources, you should have at a minimum one device with Internet access that can be displayed to the class by an LCD projector with an interactive whiteboard or a large screen arranged for class viewing. Access to a computer lab or to enough devices in your classroom for students to work in small groups is also required during one investigation, and recommended during others.

The digital resources are available online at www.FOSSweb.com for teachers and students. We recommend you access FOSSweb well in advance of starting the course to set up your teacher-user account and become familiar with the resources.

REQUIREMENTS FOR ACCESSING FOSSWEB

You'll need to have a few things in place on your device before accessing FOSSweb. Once you're online, you'll create a FOSSweb account. All information in this section is updated as needed on FOSSweb.

Creating a FOSSweb Teacher Account

By creating a FOSSweb teacher account, you can personalize FOSSweb for easy access to the courses you are teaching. When you log in, you will be able to add courses to your "My FOSS Modules" area and access Resources by Investigation for the **Electromagnetic Force Course**. This makes it simple to select the investigation and part you are teaching and view all the digital resources connected to that part.

Students and families can also access course resources through FOSSweb. You can set up a class account and class pages where students will be able to access notes from you about assignments and digital resources.

Setting up an account. Set up an account on FOSSweb so you can access the site when you begin teaching a course. Go to FOSSweb to register for an account—complete registration instructions are available online.

Entering your access code. Once your account is set up, go to FOSSweb and log in. The first time you log in, you will need to enter your access code. Your access code should be printed on the inside cover of your *Investigations Guide*. If you cannot find your FOSSweb access code, contact your school administrator, your district science coordinator, or the purchasing agent for your school or district.

Familiarize yourself with the layout of the site and the additional resources available by using your account login. From your course page, you will be able to access teacher masters, assessment masters, notebook sheets, and other digital resources. Explore the Resources by Investigation section, as this will help you plan. It lists the digital resources, notebook sheets, teacher masters, and readings for each investigation part. There are also a variety of beneficial resources on FOSSweb that can be used to assist with teacher preparation and materials management.

Setting up class pages and student accounts. To enable your students to log in to FOSSweb to access the digital resources and see class assignments, set up a class page and generate a user name and password for the class. To do so, log in to FOSSweb and go to your teacher homepage. Under My Class Pages, follow the instructions to create a new class page and to leave notes for students.

If a class page and student accounts are not set up, students can always access digital resources by visiting FOSSweb.com and choosing to visit the site as a guest.

FOSSweb Technical Requirements

To use FOSSweb, your device must meet minimum system requirements and have a compatible browser and recent versions of any required plug-ins. The system requirements are subject to change. You must visit FOSSweb to review the most recent minimum system requirements.

Preparing your browser. FOSSweb requires a supported operating system with current versions of all required plug-ins. You may need administrator privileges on your device in order to install the required programs and/or help from your school's technology coordinator. Check compatibility for each device you will use to access FOSSweb by accessing the Technical Requirements page on FOSSweb. The information on FOSSweb contains the most up-to-date technical requirements.

https://www.FOSSweb.com/tech-specs-and-info

Support for plug-ins and reader. Any required Adobe plug-ins are available on www.Adobe.com as free downloads. If required, QuickTime is available free of charge from www.Apple.com. FOSS does not support these programs. Please go to the program's website for troubleshooting information.

Accessing FOSS Electromagnetic Force Digital Resources

When you log in to FOSSweb, the most useful way to access course materials on a daily basis is the Resources by Investigation section of the **Electromagnetic Force Course** page. This section lists the digital resources, student and teacher sheets, readings, and focus questions for each investigation part. Each of these items is linked so you can click and go directly to that item.

Students will access digital resources from the Resource Room, accessible from the class page you've set up. Explore where the activities reside in the Resource Room. At various points in the course, students will access interactive simulations, images, videos, and animations from FOSSweb.

Other Technology Considerations

Firewall or proxy settings. If your school has a firewall or proxy server, contact your IT administrator to add explicit exceptions in your proxy server and firewall for these servers:

- fossweb.com
- fossweb.schoolspecialty.com
- archive.fossweb.com
- science.video.schoolspecialty.com
- a445.w10.akamai.net

Classroom technology setup. FOSS has a number of digital resources and makes every effort to accommodate users with different levels of access to technology. The digital resources can be used in a variety of ways and can be adapted to a number of classroom setups.

Teachers with classroom devices and an LCD projector, an interactive whiteboard, or a large screen will be able to show multimedia to the class. If you have access to a computer lab, or enough devices in your classroom for students to work in small groups, you can set up time for students to use the FOSSweb digital resources during the school day.

Displaying digital content. You might want to digitally display the notebook and teacher masters during class. In the Resources by Investigation section of FOSSweb, you'll have the option of downloading the masters "to project" or "to copy." Choose "to project" if you plan on projecting the masters to the class. These masters are optimized for a projection system and allow you to type into them while they are displayed. The "to copy" versions are sized to minimize paper use when photocopying for the class, and to fit optimally into student notebooks.

If this projection technology is not available to you, consider making transparencies of the notebook and teacher masters for use with an overhead projector when the Getting Ready section indicates a need to project these sheets.

> ▶ **NOTE**
> FOSSweb activities are designed for a minimum screen size of 1024 × 768. It is recommended that you adjust your screen resolution to 1024 × 768 or higher.

TROUBLESHOOTING AND TECHNICAL SUPPORT

If you experience trouble with FOSSweb, you can troubleshoot in a variety of ways.

1. Visit FOSSweb and make sure your devices meet the minimum system requirements.

    https://www.FOSSweb.com/tech-specs-and-info

2. Check the FAQs on FOSSweb for additional information that may help resolve the problem.

3. Try emptying the cache from your browser and/or quitting and relaunching it.

4. Restart your device, and make sure all hardware turns on and is connected correctly.

5. If your school has a firewall or proxy server, contact your IT administrator to add explicit exceptions in your proxy server and firewall for the servers listed on the previous page in this chapter.

If you are still experiencing problems after taking these steps, send FOSS Technical Support an e-mail at support@FOSSweb.com. In addition to describing the problem, include the following information about your device: name of device, operating system, browser and version, plug-ins and versions. This will help us troubleshoot the problem.

Science Notebook Masters

Pushes and Pulls A

Part 1: Pushing and pulling different masses

You will need one spring scale, three loads, a rubber band, and a plastic mat. Use a rubber band to bind three loads together before pulling them.

1. How much force in newtons does it take to **push** these masses?

	Trial 1	Trial 2	Trial 3	Average
1 load				
3 loads				

Predict the force required to push 2 loads. _____

What force was needed to push 2 loads? _____

2. How much force in newtons does it take to **pull** these masses?

	Trial 1	Trial 2	Trial 3	Average
1 load				
3 loads				

Predict the force required to pull 2 loads. _____

What force was needed to pull 2 loads? _____

3. What is the relationship between the mass of an object and the force needed to slide it across a surface?

FOSS Next Generation
© The Regents of the University of California
Can be duplicated for classroom or workshop use.

Electromagnetic Force Course
Investigation 1: What Is Force?
No. 1—Notebook Master

Pushes and Pulls A

Part 1: Pushing and pulling different masses

You will need one spring scale, three loads, a rubber band, and a plastic mat. Use a rubber band to bind three loads together before pulling them.

1. How much force in newtons does it take to **push** these masses?

	Trial 1	Trial 2	Trial 3	Average
1 load				
3 loads				

Predict the force required to push 2 loads. _____

What force was needed to push 2 loads? _____

2. How much force in newtons does it take to **pull** these masses?

	Trial 1	Trial 2	Trial 3	Average
1 load				
3 loads				

Predict the force required to pull 2 loads. _____

What force was needed to pull 2 loads? _____

3. What is the relationship between the mass of an object and the force needed to slide it across a surface?

FOSS Next Generation
© The Regents of the University of California
Can be duplicated for classroom or workshop use.

Electromagnetic Force Course
Investigation 1: What Is Force?
No. 1—Notebook Master

Pushes and Pulls B

Part 2: Push against push

You will need two spring scales.

4. What happens when spring scale A and spring scale B **both push** with a 4 N force on each other?

Spring scale A

4 N →

4 N ↓

Spring scale B

5. If you hold spring scale A still and push with a 4 N force with spring scale B, predict what the scale reading on spring scale A will be.

Prediction: _____ N

A

Hold still

4 N ↓

B

a. Measured force on spring scale A: _____
Was your prediction correct?

b. Explain why that happened.

FOSS Next Generation
© The Regents of the University of California
Can be duplicated for classroom or workshop use.

Electromagnetic Force Course
Investigation 1: What Is Force?
No. 2—Notebook Master

Pushes and Pulls B

Part 2: Push against push

You will need two spring scales.

4. What happens when spring scale A and spring scale B **both push** with a 4 N force on each other?

Spring scale A

4 N →

4 N ↓

Spring scale B

5. If you hold spring scale A still and push with a 4 N force with spring scale B, predict what the scale reading on spring scale A will be.

Prediction: _____ N

A

Hold still

4 N ↓

B

a. Measured force on spring scale A: _____
Was your prediction correct?

b. Explain why that happened.

FOSS Next Generation
© The Regents of the University of California
Can be duplicated for classroom or workshop use.

Electromagnetic Force Course
Investigation 1: What Is Force?
No. 2—Notebook Master

Pushes and Pulls C

Part 3: Forces on cars

You will need two spring scales and one car.

6. What happens when spring scale A **pushes** with a 2 N force on one side of the car and spring scale B **pushes** with a 3 N force on the other side of the car?

Spring scale A 2 N → 3 N **Spring scale B**

7. What happens when A **pulls** with a 5 N force on one side of the car and B **pulls** with a 5 N force on the opposite side of a car?

A 5 N ← 5 N → B

8. Apply a 2 N **pull** with A and a 2 N **push** with B on the car.

A 2 N ↑ ↑ 2 N B

a. Explain what happens to the car when forces are applied.

b. How could you use one pusher to achieve the same result?

9. What causes an object to start moving?

FOSS Next Generation
© The Regents of the University of California
Can be duplicated for classroom or workshop use.

Electromagnetic Force Course
Investigation 1: What Is Force?
No. 3—Notebook Master

Pushes and Pulls C

Part 3: Forces on cars

You will need two spring scales and one car.

6. What happens when spring scale A **pushes** with a 2 N force on one side of the car and spring scale B **pushes** with a 3 N force on the other side of the car?

Spring scale A 2 N → 3 N **Spring scale B**

7. What happens when A **pulls** with a 5 N force on one side of the car and B **pulls** with a 5 N force on the opposite side of a car?

A 5 N ← 5 N → B

8. Apply a 2 N **pull** with A and a 2 N **push** with B on the car.

A 2 N ↑ ↑ 2 N B

a. Explain what happens to the car when forces are applied.

b. How could you use one pusher to achieve the same result?

9. What causes an object to start moving?

FOSS Next Generation
© The Regents of the University of California
Can be duplicated for classroom or workshop use.

Electromagnetic Force Course
Investigation 1: What Is Force?
No. 3—Notebook Master

Friction Comparison

Surface we are testing: _____

Prediction: Do you think this surface will have more or less friction with the loads than the plastic mat did?

1. How much force does it take to **push** these masses?

	Trial 1	Trial 2	Trial 3	Average
1 load				
3 loads				

2. How much force does it take to **pull** these masses?
(Use a rubber band to bind three loads together before pulling them.)

	Trial 1	Trial 2	Trial 3	Average
1 load				
3 loads				

3. Compare your data with your results from the plastic mat. Does this surface have more or less friction with the loads? Explain your claim using evidence.

FOSS Next Generation
© The Regents of the University of California
Can be duplicated for classroom or workshop use.

Electromagnetic Force Course
Investigation 1: What Is Force?
No. 4—Notebook Master

Friction Comparison

Surface we are testing: _____

Prediction: Do you think this surface will have more or less friction with the loads than the plastic mat did?

1. How much force does it take to **push** these masses?

	Trial 1	Trial 2	Trial 3	Average
1 load				
3 loads				

2. How much force does it take to **pull** these masses?
(Use a rubber band to bind three loads together before pulling them.)

	Trial 1	Trial 2	Trial 3	Average
1 load				
3 loads				

3. Compare your data with your results from the plastic mat. Does this surface have more or less friction with the loads? Explain your claim using evidence.

FOSS Next Generation
© The Regents of the University of California
Can be duplicated for classroom or workshop use.

Electromagnetic Force Course
Investigation 1: What Is Force?
No. 4—Notebook Master

Forces on Carts A

1. In class, students learned that a cart will move when a force of 50 N or greater is applied. A student pushed on the cart with a force of 10 N. The cart did not move. Why not?

2. Student A pushed on the cart with a force of 500 N. Student B pushed on the other side of the cart. The cart did not move. How much force did student B apply? Explain how you know.

3. Student A pushed on the cart with a force of 400 N. Student C pushed on the other side of the cart. The cart rolled to the left. Did student C apply more, less, or the same force as student A? Explain how you know.

FOSS Next Generation
© The Regents of the University of California
Can be duplicated for classroom or workshop use.

Electromagnetic Force Course
Investigation 1: What Is Force?
No. 5—Notebook Master

Forces on Carts A

1. In class, students learned that a cart will move when a force of 50 N or greater is applied. A student pushed on the cart with a force of 10 N. The cart did not move. Why not?

2. Student A pushed on the cart with a force of 500 N. Student B pushed on the other side of the cart. The cart did not move. How much force did student B apply? Explain how you know.

3. Student A pushed on the cart with a force of 400 N. Student C pushed on the other side of the cart. The cart rolled to the left. Did student C apply more, less, or the same force as student A? Explain how you know.

FOSS Next Generation
© The Regents of the University of California
Can be duplicated for classroom or workshop use.

Electromagnetic Force Course
Investigation 1: What Is Force?
No. 5—Notebook Master

Forces on Carts B

B

4. Student B pushed on a cart against the wall with 500 N of force. The cart didn't move. How do you explain what happened?

C   **A**   **B**

5. Student C pushed on the cart with 400 N. Student A pushed on the cart with 1,000 N of force. Student B pulled on a rope attached to the cart with 500 N of force. What will happen to the cart and why?

FOSS Next Generation
© The Regents of the University of California
Can be duplicated for classroom or workshop use.

Electromagnetic Force Course
Investigation 1: What Is Force?
No. 6—Notebook Master

Forces on Carts B

B

4. Student B pushed on a cart against the wall with 500 N of force. The cart didn't move. How do you explain what happened?

C   **A**   **B**

5. Student C pushed on the cart with 400 N. Student A pushed on the cart with 1,000 N of force. Student B pulled on a rope attached to the cart with 500 N of force. What will happen to the cart and why?

FOSS Next Generation
© The Regents of the University of California
Can be duplicated for classroom or workshop use.

Electromagnetic Force Course
Investigation 1: What Is Force?
No. 6—Notebook Master

Response Sheet—Investigation 1

A dog-sled driver prepared for her race. Her sled was loaded with equipment, but on snow and ice, the dogs moved the heavy sled. Each dog pulled with a force of 600 N.

During the race, she came to an area of grass and dirt instead of snow. The sled slowed to a stop. The driver said,

There must be something wrong with one of the dogs. One of them is pulling with a force of less than 600 N.

How would you respond to the dog-sled driver?

FOSS Next Generation
© The Regents of the University of California
Can be duplicated for classroom or workshop use.

Electromagnetic Force Course
Investigation 1: What Is Force?
No. 7—Notebook Master

Response Sheet—Investigation 1

A dog-sled driver prepared for her race. Her sled was loaded with equipment, but on snow and ice, the dogs moved the heavy sled. Each dog pulled with a force of 600 N.

During the race, she came to an area of grass and dirt instead of snow. The sled slowed to a stop. The driver said,

There must be something wrong with one of the dogs. One of them is pulling with a force of less than 600 N.

How would you respond to the dog-sled driver?

FOSS Next Generation
© The Regents of the University of California
Can be duplicated for classroom or workshop use.

Electromagnetic Force Course
Investigation 1: What Is Force?
No. 7—Notebook Master

Response Sheet—Investigation 2

If you put something like a piece of cardboard between a magnet and an iron nail, the magnet still holds the nail in place, even though the magnet is not touching the nail.

Explain how that happens. Use the word *induce* in your response.

FOSS Next Generation
© The Regents of the University of California
Can be duplicated for classroom or workshop use.

Electromagnetic Force Course
Investigation 2: The Force of Magnetism
No. 8—Notebook Master

Response Sheet—Investigation 2

If you put something like a piece of cardboard between a magnet and an iron nail, the magnet still holds the nail in place, even though the magnet is not touching the nail.

Explain how that happens. Use the word *induce* in your response.

FOSS Next Generation
© The Regents of the University of California
Can be duplicated for classroom or workshop use.

Electromagnetic Force Course
Investigation 2: The Force of Magnetism
No. 8—Notebook Master

Adding Magnets

1. Position a paper clip on the diagram below so the end is at the 0 mm mark on the ruler. Stand a magnet on edge at 50 mm.

2. Slowly slide the magnet to the left until the paper clip jumps toward the magnet.

3. Record the position of the magnet (jump distance) when the paper clip moved.

4. Collect data for three trials, then repeat with two magnets, three magnets, and four magnets.

5. Set up a data table for multiple trials with one to four magnets.

Millimeters (mm)

Conclusion

FOSS Next Generation
© The Regents of the University of California
Can be duplicated for classroom or workshop use.

Electromagnetic Force Course
Investigation 2: The Force of Magnetism
No. 9—Notebook Master

Adding Magnets

1. Position a paper clip on the diagram below so the end is at the 0 mm mark on the ruler. Stand a magnet on edge at 50 mm.

2. Slowly slide the magnet to the left until the paper clip jumps toward the magnet.

3. Record the position of the magnet (jump distance) when the paper clip moved.

4. Collect data for three trials, then repeat with two magnets, three magnets, and four magnets.

5. Set up a data table for multiple trials with one to four magnets.

Millimeters (mm)

Conclusion

FOSS Next Generation
© The Regents of the University of California
Can be duplicated for classroom or workshop use.

Electromagnetic Force Course
Investigation 2: The Force of Magnetism
No. 9—Notebook Master

Lighting Lightbulbs

Write a prediction for each circuit in the small box. If you think it will light, write "yes." If you think it won't light, write "no."

FOSS Next Generation
© The Regents of the University of California
Can be duplicated for classroom or workshop use.

Electromagnetic Force Course
Investigation 3: Electromagnetism
No. 10—Notebook Master

Lighting Lightbulbs

Write a prediction for each circuit in the small box. If you think it will light, write "yes." If you think it won't light, write "no."

FOSS Next Generation
© The Regents of the University of California
Can be duplicated for classroom or workshop use.

Electromagnetic Force Course
Investigation 3: Electromagnetism
No. 10—Notebook Master

Response Sheet—Investigation 3

Student 1 said,

Magnetic force and electric charges are similar, because opposites attract and likes repel.

Student 2 said,

Magnetism and electricity are similar because they both have a force field extending out from them so they don't have to touch to work.

How would you respond to these students?

FOSS Next Generation
© The Regents of the University of California
Can be duplicated for classroom or workshop use.

Electromagnetic Force Course
Investigation 3: Electromagnetism
No. 11—Notebook Master

Response Sheet—Investigation 3

Student 1 said,

Magnetic force and electric charges are similar, because opposites attract and likes repel.

Student 2 said,

Magnetism and electricity are similar because they both have a force field extending out from them so they don't have to touch to work.

How would you respond to these students?

FOSS Next Generation
© The Regents of the University of California
Can be duplicated for classroom or workshop use.

Electromagnetic Force Course
Investigation 3: Electromagnetism
No. 11—Notebook Master

Testing Electromagnet Variables

Variable we are testing: _____

Standard electromagnet—number of paper clips lifted			
Trial 1	Trial 2	Trial 3	Average

Modified electromagnet—number of paper clips lifted			
Trial 1	Trial 2	Trial 3	Average

Data Analysis

1. Which device was more effective, your standard or modified electromagnet? Use evidence to support your response.

Class Analysis

2. Looking at the class data, which modifications are the most effective? Use evidence to support your response.

3. Answer on a separate page: If you built an electromagnet with multiple modifications, how would you build the most powerful electromagnet?

FOSS Next Generation
© The Regents of the University of California
Can be duplicated for classroom or workshop use.

Electromagnetic Force Course
Investigation 3: Electromagnetism
No. 12—Notebook Master

Testing Electromagnet Variables

Variable we are testing: _____

Standard electromagnet—number of paper clips lifted			
Trial 1	Trial 2	Trial 3	Average

Modified electromagnet—number of paper clips lifted			
Trial 1	Trial 2	Trial 3	Average

Data Analysis

1. Which device was more effective, your standard or modified electromagnet? Use evidence to support your response.

Class Analysis

2. Looking at the class data, which modifications are the most effective? Use evidence to support your response.

3. Answer on a separate page: If you built an electromagnet with multiple modifications, how would you build the most powerful electromagnet?

FOSS Next Generation
© The Regents of the University of California
Can be duplicated for classroom or workshop use.

Electromagnetic Force Course
Investigation 3: Electromagnetism
No. 12—Notebook Master

Energy Transfers

1. What energy transfers are happening to make the motor run?

2. Compare the energy transfers of a motor to the energy transfers of a generator.

3. Compare the inner components of a battery/motor system to the inner components of a hand-crank generator.

4. What do you think will happen when we hook up batteries to the wires of the hand-crank generator?

5. What happened when you hooked up batteries to the hand-crank generator in class?

FOSS Next Generation
© The Regents of the University of California
Can be duplicated for classroom or workshop use.

Electromagnetic Force Course
Investigation 4: Energy Transfer
No. 13—Notebook Master

Energy Transfers

1. What energy transfers are happening to make the motor run?

2. Compare the energy transfers of a motor to the energy transfers of a generator.

3. Compare the inner components of a battery/motor system to the inner components of a hand-crank generator.

4. What do you think will happen when we hook up batteries to the wires of the hand-crank generator?

5. What happened when you hooked up batteries to the hand-crank generator in class?

FOSS Next Generation
© The Regents of the University of California
Can be duplicated for classroom or workshop use.

Electromagnetic Force Course
Investigation 4: Energy Transfer
No. 13—Notebook Master

Without electrical energy, just about everything in our day-to-day lives shuts down. Our reliance on electricity is a part of modern life.

Where We Get Energy

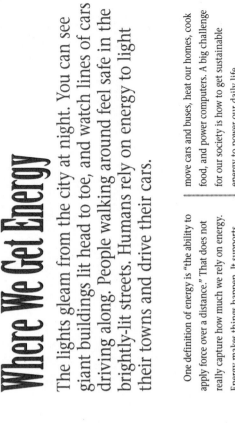

The lights gleam from the city at night. You can see giant buildings lit head to toe, and watch lines of cars driving along. People walking around feel safe in the brightly-lit streets. Humans rely on energy to light their towns and drive their cars.

One definition of energy is "the ability to apply force over a distance." That does not really capture how much we rely on energy. Energy makes things happen. It supports nearly everything we do. We use energy to move cars and buses, heat our homes, cook food, and power computers. A big challenge for our society is how to get sustainable energy to power our daily life.

(56)

FOSS Next Generation
© The Regents of the University of California
Can be duplicated for classroom or workshop use.

Electromagnetic Force Course
Investigation 4: Energy Transfer
No. 14—Notebook Master

Without electrical energy, just about everything in our day-to-day lives shuts down. Our reliance on electricity is a part of modern life.

Where We Get Energy

The lights gleam from the city at night. You can see giant buildings lit head to toe, and watch lines of cars driving along. People walking around feel safe in the brightly-lit streets. Humans rely on energy to light their towns and drive their cars.

One definition of energy is "the ability to apply force over a distance." That does not really capture how much we rely on energy. Energy makes things happen. It supports nearly everything we do. We use energy to move cars and buses, heat our homes, cook food, and power computers. A big challenge for our society is how to get sustainable energy to power our daily life.

(56)

FOSS Next Generation
© The Regents of the University of California
Can be duplicated for classroom or workshop use.

Electromagnetic Force Course
Investigation 4: Energy Transfer
No. 14—Notebook Master

The Greenhouse Effect

Some of the solar radiation passes through the atmosphere, and some is trapped in the atmosphere by greenhouse gases. The effect of this is to warm Earth's surface and the atmosphere.

Some solar radiation is reflected by clouds.

Most radiation is absorbed by Earth's surface and warms it.

Gases in the atmosphere keep our planet warm, making life possible. But the extra greenhouse gases added by global human activity are changing Earth's climate.

The Fossil-Fuel Source

Fuel is a material that stores energy. Our ancestors were burning wood as fuel more than 400,000 years ago. Humans have been burning coal as fuel for over 5,000 years. In the past 100 years or so, humans have grown used to cheap energy from fossil fuels. Fuels such as coal, oil, and natural gas have been plentiful and inexpensive. Using fossil fuels, however, creates serious problems.

When we burn fossil fuels to release energy, a chemical reaction occurs. The reaction uses the stored chemical energy of the fuel to generate thermal energy. Thermal energy can be transferred to mechanical energy and electrical energy. Mechanical energy powers the engines of cars, buses, and planes. Electrical energy travels from a power plant to our communities through the **power grid**. A power grid is a complex network of wires that delivers electricity to homes and other buildings.

Burning fossil fuels releases carbon dioxide and other pollutants into Earth's atmosphere. Carbon dioxide is a **greenhouse gas.** It helps the atmosphere "trap" thermal energy that would otherwise radiate into space.

As carbon dioxide builds up in the atmosphere, Earth becomes warmer. This is known as climate change. Human-induced climate change is driving global temperatures higher and altering climate all over Earth. Climate change leads to serious problems around the globe, like sea level rising, more severe droughts, more severe storms, and heat waves. Carbon dioxide emissions are also causing the ocean to become more acidic, endangering ocean life.

Investigation 4: Energy Transfer **57**

The Greenhouse Effect

Some of the solar radiation passes through the atmosphere, and some is trapped in the atmosphere by greenhouse gases. The effect of this is to warm Earth's surface and the atmosphere.

Some solar radiation is reflected by clouds.

Most radiation is absorbed by Earth's surface and warms it.

Gases in the atmosphere keep our planet warm, making life possible. But the extra greenhouse gases added by global human activity are changing Earth's climate.

The Fossil-Fuel Source

Fuel is a material that stores energy. Our ancestors were burning wood as fuel more than 400,000 years ago. Humans have been burning coal as fuel for over 5,000 years. In the past 100 years or so, humans have grown used to cheap energy from fossil fuels. Fuels such as coal, oil, and natural gas have been plentiful and inexpensive. Using fossil fuels, however, creates serious problems.

When we burn fossil fuels to release energy, a chemical reaction occurs. The reaction uses the stored chemical energy of the fuel to generate thermal energy. Thermal energy can be transferred to mechanical energy and electrical energy. Mechanical energy powers the engines of cars, buses, and planes. Electrical energy travels from a power plant to our communities through the **power grid**. A power grid is a complex network of wires that delivers electricity to homes and other buildings.

Burning fossil fuels releases carbon dioxide and other pollutants into Earth's atmosphere. Carbon dioxide is a **greenhouse gas.** It helps the atmosphere "trap" thermal energy that would otherwise radiate into space.

As carbon dioxide builds up in the atmosphere, Earth becomes warmer. This is known as climate change. Human-induced climate change is driving global temperatures higher and altering climate all over Earth. Climate change leads to serious problems around the globe, like sea level rising, more severe droughts, more severe storms, and heat waves. Carbon dioxide emissions are also causing the ocean to become more acidic, endangering ocean life.

Investigation 4: Energy Transfer **57**

FOSS Next Generation
© The Regents of the University of California
Can be duplicated for classroom or workshop use.

Electromagnetic Force Course
Investigation 4: Energy Transfer
No. 15—Notebook Master

Powering a Grid

Most of our electricity comes from power plants that burn fossil fuels. How does burning coal, oil, and natural gas make electricity?

Let's look at a power plant that burns coal to heat water. It creates high-temperature, high-pressure steam. That steam rushes through a **turbine** and spins it. A turbine is a machine that has many blades, like a highly engineered fan. Natural gas plants also use turbines. Instead of using steam, burning natural gas expands the gas, which spins the turbine.

The turbine turns a shaft that connects to an electric generator. The shaft is like the hand crank we used in class, only much, much larger. As the shaft rotates, coils of wire interact with permanent magnets to generate electricity. In some generators, this electricity passes through a **commutator** and **brushes**. They carry the electricity from the rotating coils to wires that connect to the power grid.

The power plant transforms the electricity for your home into a high voltage. It can travel a long way with very little energy loss in the wires.

That high-voltage electricity goes through high-voltage wires to a local transformer. The transformer lowers the voltage to 240 volts or 120 volts before delivering it to your home.

Burning fossil fuels produces about 67 percent of US electricity. Coal plants provide 33 percent, natural gas 33 percent, and petroleum about 1 percent.

Turbines are like giant pinwheels. They spin when steam, expanding gas, falling water, or wind turns their blades. A spinning turbine provides the kinetic energy for an electricity generator.

58

FOSS Next Generation
© The Regents of the University of California
Can be duplicated for classroom or workshop use.

Electromagnetic Force Course
Investigation 4: Energy Transfer
No. 16—Notebook Master

Powering a Grid

Most of our electricity comes from power plants that burn fossil fuels. How does burning coal, oil, and natural gas make electricity?

Let's look at a power plant that burns coal to heat water. It creates high-temperature, high-pressure steam. That steam rushes through a **turbine** and spins it. A turbine is a machine that has many blades, like a highly engineered fan. Natural gas plants also use turbines. Instead of using steam, burning natural gas expands the gas, which spins the turbine.

The turbine turns a shaft that connects to an electric generator. The shaft is like the hand crank we used in class, only much, much larger. As the shaft rotates, coils of wire interact with permanent magnets to generate electricity. In some generators, this electricity passes through a **commutator** and **brushes**. They carry the electricity from the rotating coils to wires that connect to the power grid.

The power plant transforms the electricity for your home into a high voltage. It can travel a long way with very little energy loss in the wires.

That high-voltage electricity goes through high-voltage wires to a local transformer. The transformer lowers the voltage to 240 volts or 120 volts before delivering it to your home.

Burning fossil fuels produces about 67 percent of US electricity. Coal plants provide 33 percent, natural gas 33 percent, and petroleum about 1 percent.

Turbines are like giant pinwheels. They spin when steam, expanding gas, falling water, or wind turns their blades. A spinning turbine provides the kinetic energy for an electricity generator.

58

FOSS Next Generation
© The Regents of the University of California
Can be duplicated for classroom or workshop use.

Electromagnetic Force Course
Investigation 4: Energy Transfer
No. 16—Notebook Master

Alternative Energy Sources

Other sources of heat are used for generating electricity. About 20 percent of US electricity comes from nuclear power. In nuclear power plants, the radioactive decay of the element uranium heats water to make steam and turn turbines. Uranium is mined and is **nonrenewable**, like fossil fuels.

Nuclear power plants do not produce greenhouse gases that would contribute to global warming. But they do produce radioactive waste. That waste must be carefully contained and stored for thousands of years. These plants present a risk to nearby communities if their equipment breaks or is damaged by a natural disaster, like an earthquake.

Fossil fuels and nuclear fuel are not sustainable energy sources. Fossil fuels form over hundreds of millions of years and cannot be replaced as fast as we are using them up. The situation with nuclear fuel is similar.

Burning wood to make steam for turbines is also possible. About 0.4 percent of US electricity is produced that way. Is wood a sustainable energy source? That depends on whether trees are replanted and how fast they can grow.

Nuclear power plants generate electricity by using steam to turn turbines. The heat to make the steam is produced not by burning fuel but by splitting the nucleus of a uranium atom, a process called fission.

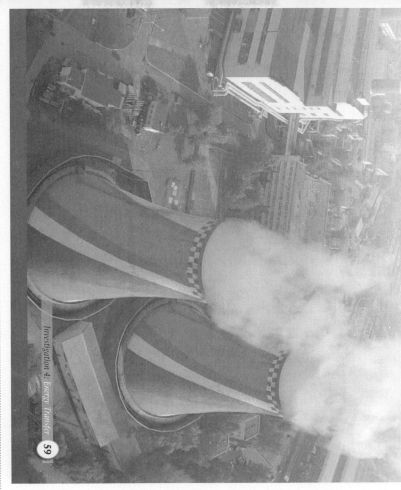

FOSS Next Generation
© The Regents of the University of California
Can be duplicated for classroom or workshop use.

Electromagnetic Force Course
Investigation 4: Energy Transfer
No. 17—Notebook Master

Alternative Energy Sources

Other sources of heat are used for generating electricity. About 20 percent of US electricity comes from nuclear power. In nuclear power plants, the radioactive decay of the element uranium heats water to make steam and turn turbines. Uranium is mined and is **nonrenewable**, like fossil fuels.

Nuclear power plants do not produce greenhouse gases that would contribute to global warming. But they do produce radioactive waste. That waste must be carefully contained and stored for thousands of years. These plants present a risk to nearby communities if their equipment breaks or is damaged by a natural disaster, like an earthquake.

Fossil fuels and nuclear fuel are not sustainable energy sources. Fossil fuels form over hundreds of millions of years and cannot be replaced as fast as we are using them up. The situation with nuclear fuel is similar.

Burning wood to make steam for turbines is also possible. About 0.4 percent of US electricity is produced that way. Is wood a sustainable energy source? That depends on whether trees are replanted and how fast they can grow.

Nuclear power plants generate electricity by using steam to turn turbines. The heat to make the steam is produced not by burning fuel but by splitting the nucleus of a uranium atom, a process called fission.

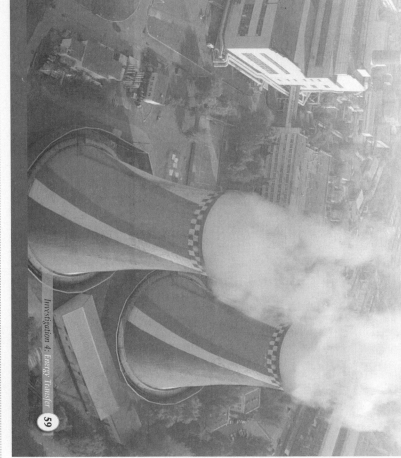

Investigation 4: Energy Transfer

59

FOSS Next Generation
© The Regents of the University of California
Can be duplicated for classroom or workshop use.

Electromagnetic Force Course
Investigation 4: Energy Transfer
No. 17—Notebook Master

Hydropower, produced by the force of flowing or falling water, is the most widely used renewable energy source. Hydroelectric dams are nonpolluting, but they do have significant environmental impacts.

Iceland has lots of geothermal activity. About 25 percent of Iceland's electricity comes from geothermal energy.

Sustainable Energy Sources

One source of sustainable energy is Earth's heat. In some places, Earth's crust is quite thin. Water just beneath the surface can be very hot. Sometimes, geological features called geysers shoot steam and water out of the ground, or hot springs form. Geothermal power plants in these locations use high-pressure steam or very hot water from deep underground to run turbines. Geothermal power plants do not produce greenhouse gases. They produce only about 0.5 percent of US electricity.

A turbine can spin without needing to burn fuel or collect heated steam. Water flowing through a dam can turn turbines.

The turbine connects to a generator to make electricity. This sustainable method, called hydroelectric power, produces no greenhouse gases. It produces about 6 percent of US electricity.

Hydroelectric power does create problems. Damming rivers and converting them into lakes has serious effects on surrounding ecosystems. As the lake behind the dam fills up with sediment, the water pressure drops. Lower water pressure reduces the efficiency of the power plant. Engineers are working to develop water turbines to harness power from rivers, ocean waves, and tides. These turbines are more efficient and do less harm to ecosystems than turbines in dams.

60

FOSS Next Generation
© The Regents of the University of California
Can be duplicated for classroom or workshop use.

Electromagnetic Force Course
Investigation 4: Energy Transfer
No. 18—Notebook Master

Hydropower, produced by the force of flowing or falling water, is the most widely used renewable energy source. Hydroelectric dams are nonpolluting, but they do have significant environmental impacts.

Iceland has lots of geothermal activity. About 25 percent of Iceland's electricity comes from geothermal energy.

Sustainable Energy Sources

One source of sustainable energy is Earth's heat. In some places, Earth's crust is quite thin. Water just beneath the surface can be very hot. Sometimes, geological features called geysers shoot steam and water out of the ground, or hot springs form. Geothermal power plants in these locations use high-pressure steam or very hot water from deep underground to run turbines. Geothermal power plants do not produce greenhouse gases. They produce only about 0.5 percent of US electricity.

A turbine can spin without needing to burn fuel or collect heated steam. Water flowing through a dam can turn turbines.

The turbine connects to a generator to make electricity. This sustainable method, called hydroelectric power, produces no greenhouse gases. It produces about 6 percent of US electricity.

Hydroelectric power does create problems. Damming rivers and converting them into lakes has serious effects on surrounding ecosystems. As the lake behind the dam fills up with sediment, the water pressure drops. Lower water pressure reduces the efficiency of the power plant. Engineers are working to develop water turbines to harness power from rivers, ocean waves, and tides. These turbines are more efficient and do less harm to ecosystems than turbines in dams.

60

FOSS Next Generation
© The Regents of the University of California
Can be duplicated for classroom or workshop use.

Electromagnetic Force Course
Investigation 4: Energy Transfer
No. 18—Notebook Master

Variable Energy Sources

Wind turbines can be connected to generators. This technology is simple, inexpensive, and sustainable. It produces about 5 percent of US electricity. One drawback of wind-electric power is that it varies, depending on wind speed. And when wind speeds are too high, wind turbines must stop to avoid damage. For this reason, wind-generated electricity is usually stored. It is just one source of electricity connected to a power grid.

Photovoltaic technology does not depend on electric generators at all. It uses solar panels made of flat arrays of **solar cells.** When sunlight shines on a solar cell, part of the energy in the light is converted directly into electricity.

You can find solar panels on the roofs of some schools, businesses, parking lots, and homes. Some very large power plants in deserts use solar cells to generate electricity for cities. Only about 0.7 percent of US

electricity comes from solar panels. Solar installations are rapidly growing, so the percentage of US electricity that is solar could increase dramatically in years to come.

Like wind power, solar power is variable. It does not generate electricity at night or on dark stormy days. This energy source is sustainable in the long run, but must be stored or added to other sources of electricity. One relatively simple way to store electricity is in rechargeable batteries. Another is using the electricity to pump water uphill. Holding the water in that position stores **potential energy.** To produce electricity, the water flows through a turbine in a hydroelectric system when the stored energy is needed as electricity again.

Take Note

Ask your local electric company what energy sources and technologies provide your community's electricity.

Variable Energy Sources

Wind turbines can be connected to generators. This technology is simple, inexpensive, and sustainable. It produces about 5 percent of US electricity. One drawback of wind-electric power is that it varies, depending on wind speed. And when wind speeds are too high, wind turbines must stop to avoid damage. For this reason, wind-generated electricity is usually stored. It is just one source of electricity connected to a power grid.

Photovoltaic technology does not depend on electric generators at all. It uses solar panels made of flat arrays of **solar cells.** When sunlight shines on a solar cell, part of the energy in the light is converted directly into electricity.

You can find solar panels on the roofs of some schools, businesses, parking lots, and homes. Some very large power plants in deserts use solar cells to generate electricity for cities. Only about 0.7 percent of US

electricity comes from solar panels. Solar installations are rapidly growing, so the percentage of US electricity that is solar could increase dramatically in years to come.

Like wind power, solar power is variable. It does not generate electricity at night or on dark stormy days. This energy source is sustainable in the long run, but must be stored or added to other sources of electricity. One relatively simple way to store electricity is in rechargeable batteries. Another is using the electricity to pump water uphill. Holding the water in that position stores **potential energy.** To produce electricity, the water flows through a turbine in a hydroelectric system when the stored energy is needed as electricity again.

Take Note

Ask your local electric company what energy sources and technologies provide your community's electricity.

Investigation 4: Energy Transfer   **61**

Electromagnetic Force Course
Investigation 4: Energy Transfer
No. 19—Notebook Master

FOSS Next Generation
© The Regents of the University of California
Can be duplicated for classroom or workshop use.

The Future of Renewable Energy

As we start to experience the effects of climate change, countries around the world are working together to reduce carbon dioxide emissions. Renewable energy sources supply just 13 percent of US power, but it's a start. What are other countries doing?

- Sweden is becoming 100 percent fossil-fuel free. It is investing heavily in solar and wind power and energy storage.
- Germany has turned its attention to solar energy. Depending on weather conditions and time of year, it produces up to 78 percent of its electricity from renewable sources.
- China is committed to phasing out coal burning. It is also addressing its serious air pollution problems. It is the largest producer of renewable energy in the world, mostly from wind and solar power.

The United Nations brings countries together to use more sustainable energy sources. Each year, countries agree to goals that each country works to meet. These goals call on engineers to develop new technologies. They encourage all people to reduce energy use. We can all help reduce carbon emissions and try to slow climate change.

Solar energy is the cleanest and most abundant energy source available. These rooftop panels convert sunlight to electricity that can be stored and used to meet household power needs.

Think Questions

1. Why is using fossil fuels considered unsustainable?
2. What are some advantages of using sustainable energy sources?
3. Compare the different energy sources described in this article. In your notebook, draw a table like the one below. Mark with an X to show how each energy source works. The first row shows coal as an example.

Energy source	Uses steam	Uses turbine	Uses generator	Emits carbon dioxide	Sustainable in long term
Coal	X	X	X	X	

FOSS Next Generation
© The Regents of the University of California
Can be duplicated for classroom or workshop use.

Electromagnetic Force Course
Investigation 4: Energy Transfer
No. 20—Notebook Master

The Future of Renewable Energy

As we start to experience the effects of climate change, countries around the world are working together to reduce carbon dioxide emissions. Renewable energy sources supply just 13 percent of US power, but it's a start. What are other countries doing?

- Sweden is becoming 100 percent fossil-fuel free. It is investing heavily in solar and wind power and energy storage.
- Germany has turned its attention to solar energy. Depending on weather conditions and time of year, it produces up to 78 percent of its electricity from renewable sources.
- China is committed to phasing out coal burning. It is also addressing its serious air pollution problems. It is the largest producer of renewable energy in the world, mostly from wind and solar power.

The United Nations brings countries together to use more sustainable energy sources. Each year, countries agree to goals that each country works to meet. These goals call on engineers to develop new technologies. They encourage all people to reduce energy use. We can all help reduce carbon emissions and try to slow climate change.

Solar energy is the cleanest and most abundant energy source available. These rooftop panels convert sunlight to electricity that can be stored and used to meet household power needs.

Think Questions

1. Why is using fossil fuels considered unsustainable?
2. What are some advantages of using sustainable energy sources?
3. Compare the different energy sources described in this article. In your notebook, draw a table like the one below. Mark with an X to show how each energy source works. The first row shows coal as an example.

Energy source	Uses steam	Uses turbine	Uses generator	Emits carbon dioxide	Sustainable in long term
Coal	X	X	X	X	

62

FOSS Next Generation
© The Regents of the University of California
Can be duplicated for classroom or workshop use.

Electromagnetic Force Course
Investigation 4: Energy Transfer
No. 20—Notebook Master

Teacher Masters

FRICTION COMPARISON DATA

Group	Surface tested	More or less friction?
1		
2		
3		
4		
5		
6		
7		
8		

FOSS Next Generation
© The Regents of the University of California
Can be duplicated for classroom or workshop use.

Electromagnetic Force Course
Investigation 1: What Is Force?
Teacher Master A

MAGNETIC FIELDS

You can detect magnetic fields around magnets using iron filings.

© iStockphoto/Tommounsey

We can represent magnetic fields as lines that connect the north and south poles of any magnet . . .

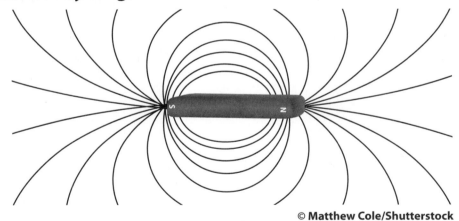

© Matthew Cole/Shutterstock

including Earth!

© Matthew Cole/Shutterstock; © worker/shutterstock; © ARTEKI/Shutterstock

FOSS Next Generation
© The Regents of the University of California
Can be duplicated for classroom or workshop use.

Electromagnetic Force Course
Investigation 2: The Force of Magnetism
Teacher Master B

MAGNETIC-FORCE SETUP

1. Place one magnet in a cup, then place another cup on top of it.

2. Test the magnet–on–a–post to see if it is attracted to the magnet in the cup by placing it under the cup. If it repels, flip the cup magnet to face the other way.

3. Attach the spring scale to the loop and pull slowly and gently to measure the force (N) required to pull apart the magnets.

FOSS Next Generation
© The Regents of the University of California
Can be duplicated for classroom or workshop use.

Electromagnetic Force Course
Investigation 2: The Force of Magnetism
Teacher Master C

USING SPACERS

When you are ready to test with spacers, arrange the spacers in the bottom cup so they will be *between* the magnets.

FOSS Next Generation
© The Regents of the University of California
Can be duplicated for classroom or workshop use.

Electromagnetic Force Course
Investigation 2: The Force of Magnetism
Teacher Master D

INSIDE AN INCANDESCENT LIGHTBULB

Filament

Glass globe

Filament support wire

Casing wire →

Metal casing

Base

Base insulator

Base wire

© iStockphoto/heinrikroger

FOSS Next Generation
© The Regents of the University of California
Can be duplicated for classroom or workshop use.

Electromagnetic Force Course
Investigation 3: Electromagnetism
Teacher Master E

NEW CIRCUIT CHALLENGE

Each group will need:

    2 D-cells

    2 cell holders

    2 lightbulbs

    2 bulb holders

    2 switches

FOSS Next Generation
© The Regents of the University of California
Can be duplicated for classroom or workshop use.

Electromagnetic Force Course
Investigation 3: Electromagnetism
Teacher Master F

ELECTROMAGNET SETUP

A. Wrap the thin wire 40 times around the rivet. Be sure to leave about 10 cm of wire free at the start end so you can attach it to a D-cell.

B. Make sure all wraps are in the same direction. Keep the wraps between the head of the rivet and the first black washer.

C. Put the rivet/wire system into a circuit, using one D-cell in a holder and a switch.

D. When connecting wires to the D-cell, connect to the spring closest to the terminal for the best results.

E. Always leave the switch open unless you are testing the rivet, or you might drain the D-cells.

F. See if you can use the rivet as a magnet by testing it with paper clips.

FOSS Next Generation
© The Regents of the University of California
Can be duplicated for classroom or workshop use.

Electromagnetic Force Course
Investigation 3: Electromagnetism
Teacher Master G

ELECTROMAGNET VARIABLES—CLASS DATA

Variable tested	Magnet strengthened or weakened?	Difference in average paper clips

DISSECTED MOTOR IN THREE PARTS

Casing   **Shaft**   **Cap**

FOSS Next Generation
© The Regents of the University of California
Can be duplicated for classroom or workshop use.

Electromagnetic Force Course
Investigation 4: Energy Transfer
Teacher Master I

MOTOR SHAFT AND COMMUTATOR

Gap in commutator

Shaft

Commutator

Electromagnet sections

MOTOR CAP AND BRUSHES

Brushes   **Cap**

FOSS Next Generation
© The Regents of the University of California
Can be duplicated for classroom or workshop use.

Electromagnetic Force Course
Investigation 4: Energy Transfer
Teacher Master K

HAND-CRANK GENERATOR ....

Assessment Masters

Embedded Assessment Notes

Electromagnetic Force

Investigation ____, Part ____   Date _____

Concept:

Tally: _____ Got it _____ | _____ Doesn't get it _____

Misconceptions/incomplete ideas:

Reflections/next steps:

Investigation ____, Part ____   Date _____

Concept:

Tally: _____ Got it _____ | _____ Doesn't get it _____

Misconceptions/incomplete ideas:

Reflections/next steps:

Investigation ____, Part ____   Date _____

Concept:

Tally: _____ Got it _____ | _____ Doesn't get it _____

Misconceptions/incomplete ideas:

Reflections/next steps:

FOSS Next Generation
© The Regents of the University of California
Can be duplicated for classroom or workshop use.

Electromagnetic Force Course
Embedded Assessment Notes
No. 1—Assessment Master

Performance Assessment Checklist by Group

Electromagnetic Force

Investigation 1, Part 2

Group	Science and Engineering Practices			DCI	Crosscutting Concept
	Planning and carrying out investigations	Analyzing and interpreting data	Using mathematics and computational thinking	PS2.A Force and motion	Cause and effect

FOSS Next Generation
© The Regents of the University of California
Can be duplicated for classroom or workshop use.

NOTE: See the Assessment chapter for a discussion about how to use this checklist.

Performance Assessment Checklist by Student

Electromagnetic Force

Investigation 1, Part 2

Student	Science and Engineering Practices			DCI	Crosscutting Concept
	Planning and carrying out investigations	Analyzing and interpreting data	Using mathematics and computational thinking	PS2.A Force and motion	Cause and effect

NOTE: See the Assessment chapter for a discussion about how to use this checklist.

Performance Assessment Checklist by Group

Electromagnetic Force

Investigation 2, Part 3

Group	Science and Engineering Practices				DCI	Crosscutting Concepts	
	Developing and using models	Planning and carrying out investigations	Analyzing and interpreting data	Using mathematics and computational thinking	PS2.B Types of interactions	Patterns	Systems and system models

NOTE: See the Assessment chapter for a discussion about how to use this checklist.

FOSS Next Generation
© The Regents of the University of California
Can be duplicated for classroom or workshop use.

Electromagnetic Force Course
Performance Assessment Checklist
No. 4—Assessment Master

Performance Assessment Checklist by Student

Electromagnetic Force

Investigation 2, Part 3

Student	Science and Engineering Practices				DCI	Crosscutting Concepts	
	Developing and using models	Planning and carrying out investigations	Analyzing and interpreting data	Using mathematics and computational thinking	PS2.B Types of interactions	Patterns	Systems and system models

NOTE: See the Assessment chapter for a discussion about how to use this checklist.

Performance Assessment Checklist by Group

Electromagnetic Force

Investigation 3, Part 3

| Group | Science and Engineering Practices | | | | | DCI | Crosscutting Concepts | |
	Planning and carrying out investigations	Analyzing and interpreting data	Using mathematics & computational thinking	Designing solutions	Engaging in argument from evidence	ETS1.C Optimizing the design solution	Cause and effect	Systems and system models

FOSS Next Generation
© The Regents of the University of California
Can be duplicated for classroom or workshop use.

NOTE: See the Assessment chapter for a discussion about how to use this checklist.

Performance Assessment Checklist by Student

Investigation 3, Part 3

Electromagnetic Force

Student	Science and Engineering Practices					DCI	Crosscutting Concepts	
	Planning and carrying out investigations	Analyzing and interpreting data	Using mathematics & computational thinking	Designing solutions	Engaging in argument from evidence	ETS1.C Optimizing the design solution	Cause and effect	Systems and system models

NOTE: See the Assessment chapter for a discussion about how to use this checklist.

Performance Assessment Checklist by Group

Electromagnetic Force

Investigation 4, Part 2

| Group | Science and Engineering Practices | | | DCI | Crosscutting Concepts | | |
	Planning and carrying out investigations	Analyzing and interpreting data	Constructing explanations	PS3.B Conservation of energy and energy transfer	Cause and effect	Systems and system models	Energy and matter

NOTE: See the Assessment chapter for a discussion about how to use this checklist.

FOSS Next Generation
© The Regents of the University of California
Can be duplicated for classroom or workshop use.

Electromagnetic Force Course
Performance Assessment Checklist
No. 8—Assessment Master

Performance Assessment Checklist by Student

Investigation 4, Part 2

Student	Science and Engineering Practices			DCI	Crosscutting Concepts		
	Planning and carrying out investigations	Analyzing and interpreting data	Constructing explanations	PS3.B Conservation of energy and energy transfer	Cause and effect	Systems and system models	Energy and matter

NOTE: See the Assessment chapter for a discussion about how to use this checklist.

Assessment Record—Entry-Level Survey

Electromagnetic Force

Item	Contributes to	Notes for Planning Instruction
1	MS-PS2-2	
2	MS-PS2-2	
3	MS-PS2-3 MS-PS2-5	
4	MS-PS3-2 MS-PS3-5	
5	MS-PS2-3	
6	MS-PS3-5	

Assessment Record—Investigation 1 I-Check

Electromagnetic Force

Student names	1	2a	2b	3	4	5	6	7	8

NOTE: A spreadsheet for this chart is available on FOSSweb.com

Assessment Record—Investigation 2 I-Check

Electromagnetic Force

Student names	1	2	3	4	5	6	7a	7b	8	9

NOTE: A spreadsheet for this chart is available on FOSSweb.com

FOSS Next Generation
© The Regents of the University of California
Can be duplicated for classroom or workshop use.

Electromagnetic Force Course
Assessment Record
No. 12—Assessment Master

Assessment Record—Investigation 3 I-Check

Electromagnetic Force

Student names	1	2	3a	3b	4	5	6	7

NOTE: A spreadsheet for this chart is available on FOSSweb.com

Assessment Record—Investigation 4 I-Check

Electromagnetic Force

Student names	1	2	3	4	5

NOTE: A spreadsheet for this chart is available on FOSSweb.com

Assessment Record—Posttest

Electromagnetic Force

Student names	1	2	3	4	5	6	7	8	9	10	11	12	13	14

NOTE: A spreadsheet for this chart is available on FOSSweb.com

ENTRY-LEVEL SURVEY
ELECTROMAGNETIC FORCE

Name _____

Date _____ Class _____

1. What makes things move?

_____

_____

_____

_____

_____

_____

_____

2. A soccer ball is sitting on the ground.

What, if any, forces are acting on the ball when it is sitting on the ground?

_____

_____

_____

A player kicks the ball, and it rolls along the ground. What, if any, forces are acting on the ball when it is kicked, rolls, and finally stops?

_____

_____

_____

FOSS Next Generation
© The Regents of the University of California
Can be duplicated for classroom or workshop use.

Electromagnetic Force Course
Entry-Level Survey
Page 1 of 3

System A   System B

3. Study the pictures of magnets on a post.

Explain the forces that are being applied in System A.

_____

_____

_____

Explain the forces that are being applied in System B.

_____

_____

_____

4. A student hooked up a solar panel to a small motor and a switch. He placed the system in the sunlight. When the switch was closed, the motor began to turn.

Draw a diagram of the circuit this student built. Then describe the energy source and how energy transfers to make the motor move.

_____

_____

_____

_____

_____

_____

_____

_____

_____

_____

FOSS Next Generation
© The Regents of the University of California
Can be duplicated for classroom or workshop use.

Electromagnetic Force Course
Entry-Level Survey
Page 2 of 3

5. What is the relationship between electricity and magnetism?

_____

_____

_____

_____

_____

_____

_____

_____

6. A student connects a battery to a circuit with a motor. Explain where potential and kinetic energy occur in this system.

_____

_____

_____

_____

_____

_____

_____

_____

_____

_____

FOSS Next Generation
© The Regents of the University of California
Can be duplicated for classroom or workshop use.

Electromagnetic Force Course
Entry-Level Survey
Page 3 of 3

INVESTIGATION 1 I-CHECK
ELECTROMAGNETIC FORCE

1. A plant is sitting on a table.

    What, if any, forces are acting on the plant? (If there are forces acting on the plant, include what direction they are acting.)

    _____

    _____

    _____

    _____

    _____

2. A student pushes a cart against a wall exerting 500 N of force. The cart doesn't move.

    a. How do you explain what happened?
        (Mark the one best answer.)

        ○ **A**   The wall pushes back with 1,000 N of force.

        ○ **B**   The wall pushes back with 250 N of force.

        ○ **C**   The wall pushes back with 500 N of force.

        ○ **D**   Walls are stationary; they can't push or pull.

    b. What can you say about the forces?

        (Mark the one best answer.)

        ○ **F**   The net force is 0 because the cart is not moving.

        ○ **G**   The net force is greater than 0 because the wall could push back more.

        ○ **H**   There is no net force because a wall can't push or pull.

        ○ **J**   The net force is less than 0 because the person is pushing on the wall.

FOSS Next Generation
© The Regents of the University of California
Can be duplicated for classroom or workshop use.

Electromagnetic Force Course
Investigation 1 I-Check
Page 1 of 4

INVESTIGATION 1 I-CHECK
ELECTROMAGNETIC FORCE

Team A   Team B

3. Team A pulls on a rope in a game of tug-of-war with 1,800 N of force. Team B pulls on the other side of the rope. In between the teams is a mud pit. How much force does Team B need to use in order to pull Team A into the mud?

_____

4. Three young children are fighting over where to take Super Dog. Child X pulls the wagon, trying to go in the direction of his house. Child Y pushes on the wagon toward the same house. Child Z pulls on the wagon, trying to move it in the opposite direction. The forces from the three children being exerted on the wagon are 300 N, 500 N, and 200 N (not necessarily in XYZ order).

If the wagon isn't moving, how much force is Child Z exerting on the wagon?

(Mark the one best answer.)

○ **A** There is not enough information to determine Student Z's force.

○ **B** 200 N

○ **C** 300 N

○ **D** 500 N

Y pushes   X pulls

Z pushes

FOSS Next Generation
© The Regents of the University of California
Can be duplicated for classroom or workshop use.

Electromagnetic Force Course
Investigation 1 I-Check
Page 2 of 4

INVESTIGATION 1 I-CHECK
ELECTROMAGNETIC FORCE
• •

5. Write **0** next to each statement that is an example of a situation where net force is 0 N; write **X** if the net force is more or less than 0 N.

_____ Two people pushing equally on the same side of a couch moving across the floor

_____ Two people pushing a couch equally from opposite sides

_____ Two people lifting a couch

_____ A couch sliding across a frozen lake

_____ A couch coming to a stop on the frozen lake

_____ A couch sitting still on the frozen lake

_____ A couch falling through the ice

6. Read each statement about forces. Write **T** if the statement is true; write **F** if the statement is false.

_____ A force is always a push.

_____ Rougher surfaces create more friction, so more force is needed to move something over the rougher surface.

_____ It takes more force to move an object with less mass.

_____ To stop an object that is in motion on a frictionless surface, you stop applying a force.

_____ Force is always an interaction between two objects.

_____ When a force sets an object in motion, the moving object has kinetic energy.

FOSS Next Generation
© The Regents of the University of California
Can be duplicated for classroom or workshop use.

Electromagnetic Force Course
Investigation 1 I-Check
Page 3 of 4

7. In order to move a carton of books across table A, a student had to push with 15 N of force. In order to move a carton of books with the same mass across table B, the student had to push with 25 N of force.

Which table has a smoother surface?

○ **A** Table A

○ **B** Table B

What is your evidence?

_____

_____

_____

8. Aristotle had ideas developed from thought experiments about force that people believed for more than 2,000 years. In the 1600s, Galileo conducted investigations with objects and presented new ideas about force. Which great thinker had better evidence to support his claims?

(Mark the one best answer.)

○ **A** Aristotle, because he was the first to describe the motion of objects

○ **B** Aristotle, because he used thought experiments to support his ideas

○ **C** Galileo, because he used ball-and-ramp experiments to support his ideas

○ **D** Galileo, because he was a scientist rather than a philosopher

FOSS Next Generation
© The Regents of the University of California
Can be duplicated for classroom or workshop use.

Electromagnetic Force Course
Investigation 1 I-Check
Page 4 of 4

INVESTIGATION 2 I-CHECK
ELECTROMAGNETIC FORCE

1. If you cut a permanent magnet into four pieces, you will have _____.

    (Mark the one best answer.)

    ○ **A** four temporary magnets

    ○ **B** four smaller permanent magnets

    ○ **C** two north poles and two south poles

    ○ **D** four objects that are no longer magnets

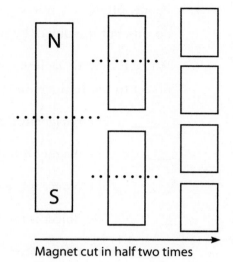

Magnet cut in half two times

2. A magnet exerts a force, but you can't see it.

    Write **T** next to each example that provides evidence that this is true; write **F** if the example does not provide the needed evidence.

    _____ A paper clip moves toward a magnet when the two objects are about 2 cm apart.

    _____ Sometimes you cannot push two magnets together to make them touch.

    _____ A magnet is placed on an aluminum pan; when you hold the magnet only, the pan falls to the floor.

    _____ Iron filings on a plate move into a distinct pattern when gently shaken above a magnet.

3. Fill in each blank with the correct term: **potential** or **kinetic**.

    When a magnet causes a paper clip to move, it's an example of _____ energy being changed into _____ energy.

    A child running is an example of _____ energy.

    A motionless iron object in the magnetic field of a magnet may be said to have _____ energy.

FOSS Next Generation
© The Regents of the University of California
Can be duplicated for classroom or workshop use.

Electromagnetic Force Course
Investigation 2 I-Check
Page 1 of 4

INVESTIGATION 2 I-CHECK
ELECTROMAGNETIC FORCE

4. People often stick notes and other objects to refrigerator doors, using magnets. A student tried to do this, but the object he wanted to stick to the refrigerator fell to the floor.

    Write **E** next to each statement that helps explain why the magnet was not able to hold the object to the refrigerator door; write **X** if the statement does not help explain what happened.

    _____ The refrigerator door is made of steel (that contains iron).

    _____ The magnet is too weak.

    _____ The object is too thick.

    _____ The object is too light.

    _____ The refrigerator door is made of aluminum.

5. Write **T** if the sentence is true; write **F** if the sentence is false.

    _____ Magnets stick to all metals.

    _____ Two magnets can attract or repel.

    _____ A magnet has a north pole or a south pole, but not both.

    _____ A magnet has an invisible magnetic field extending from its poles.

6. The ancient Romans knew about an earth material called lodestone. Lodestone is a rock that is a natural magnet. The Romans created a monument in one of their temples that included a statue suspended in midair. How might they have used lodestone to create this effect?

    _____

    _____

    _____

    _____

    _____

    _____

FOSS Next Generation
© The Regents of the University of California
Can be duplicated for classroom or workshop use.

Electromagnetic Force Course
Investigation 2 I-Check
Page 2 of 4

Number of magnets	Number of steel washers picked up		
	Trial 1	Trial 2	Trial 3
1	8	9	7
2	13	16	14
3	22	23	24

7. Study the data table.

   a. What question was this student testing?

   (Mark the one best answer.)

   ○ **A** How many washers can you pick up with a magnet?

   ○ **B** What is the average number of washers picked up by a magnet?

   ○ **C** How does the number of magnets affect the number of washers picked up?

   ○ **D** How many trials are needed to pick up the most washers?

   b. What cause-and-effect relationship can be stated from the data in the table?

   (Mark the one best answer.)

   ○ **F** You can't say because something is wrong with the data; the student should have picked up the same number of washers for each trial.

   ○ **G** The average number of washers picked up decreases as more magnets are used.

   ○ **H** The average number of washers one magnet can pick up is 9.

   ○ **J** The more magnets you use, the more washers you can pick up.

FOSS Next Generation
© The Regents of the University of California
Can be duplicated for classroom or workshop use.

Electromagnetic Force Course
Investigation 2 I-Check
Page 3 of 4

8. Explain the difference between a temporary magnet and a permanent magnet. Use examples in your explanation.

_____

_____

_____

_____

_____

_____

_____

9. A student has a magnet and a paper clip.

Trial 1: She holds the magnet 2 cm away from the paper clip and watches the paper clip jump to the magnet.

Trial 2: She holds the magnet 10 cm away from the paper clip but the paper clip does not move.

Draw two diagrams in the boxes below to show what happened in each trial. Be sure to represent the magnetic field in some way in each of your diagrams.

A	B

FOSS Next Generation
© The Regents of the University of California
Can be duplicated for classroom or workshop use.

Electromagnetic Force Course
Investigation 2 I-Check
Page 4 of 4

INVESTIGATION 3 I-CHECK
ELECTROMAGNETIC FORCE

1. Study the diagram of an electromagnet. Identify what each component does in the circuit.

    _____ Has potential energy to run the circuit

    _____ Controls the flow of electricity by opening and closing the circuit

    _____ Becomes a temporary magnet

    _____ Provides the pathway to carry the electricity from component to component

2. You can't see electricity when it is flowing. You need to have some other kind of evidence.

    Write **E** next to each statement that describes evidence that electricity is flowing; write **X** if the statement does not provide that evidence.

    _____ A component gets hot.

    _____ A motor in a circuit is not moving.

    _____ The switch is closed.

    _____ A lightbulb shines.

FOSS Next Generation
© The Regents of the University of California
Can be duplicated for classroom or workshop use.

Electromagnetic Force Course
Investigation 3 I-Check
Page 1 of 4

INVESTIGATION 3 I-CHECK
ELECTROMAGNETIC FORCE

3. a. Which circuit will light the bulb?

○ **A**   ○ **B**   ○ **C**   ○ **D**

b. What is the energy source in the circuits shown above?

(Mark the one best answer.)

○ **F**   The bulb filament

○ **G**   The bulb casing

○ **H**   The wire

○ **J**   The D-cell (battery)

4. A student set a compass on top of a wire not yet connected in a circuit. When she connected the wire to the positive side of the D-cell, the compass needle moved.

Why did the compass needle change position when the circuit was closed?

(Mark the one best answer.)

○ **A**   When the circuit is complete, electricity also runs through the compass.

○ **B**   A magnetic field surrounds a wire when electric current flows through it.

○ **C**   The compass-needle change indicates the potential energy in the D-cell.

○ **D**   The magnetic poles of the D-cell reversed, causing the compass needle to move.

FOSS Next Generation
© The Regents of the University of California
Can be duplicated for classroom or workshop use.

Electromagnetic Force Course
Investigation 3 I-Check
Page 2 of 4

INVESTIGATION 3 I-CHECK
ELECTROMAGNETIC FORCE

5. Write **Y** (for yes) next to each statement that describes where electricity flows in an electromagnet; write **N** (for no) if the statement does not describe where electricity flows.

_____ Electricity flows through the rivet (core).

_____ Electricity flows through the plastic insulation on the wire.

_____ Electricity flows through the switch when it is closed.

_____ Electricity flows through the wire wrapped around the core.

6. The graph shows the results of an investigation with an electromagnet.

What pattern does the data show?

(Mark the one best answer.)

○ **A**   The thinner the wire, the fewer nails lifted.

○ **B**   The thicker the wire, the fewer nails lifted.

○ **C**   The thinner the wire, the fewer washers lifted.

○ **D**   The gauge of the wire does not affect how many nails can be picked up.

FOSS Next Generation
© The Regents of the University of California
Can be duplicated for classroom or workshop use.

Electromagnetic Force Course
Investigation 3 I-Check
Page 3 of 4

INVESTIGATION 3 I-CHECK
ELECTROMAGNETIC FORCE

7. A group of students from another class wanted to improve their electromagnet design. Starting with the standard design that everyone used in class, they decreased the number of winds on the core and added three more batteries. Then they tested their new design. Their electromagnet could pick up more paper clips than the original design. They recorded the following conclusion:

> The fact that our modified electromagnet picked up more paper clips tells us that less winds and more batteries are both ways to make an electromagnet stronger.

Evaluate this group's investigation and explain to them why you agree or why you disagree with their conclusion.

_____

_____

_____

_____

_____

_____

_____

_____

FOSS Next Generation
© The Regents of the University of California
Can be duplicated for classroom or workshop use.

Electromagnetic Force Course
Investigation 3 I-Check
Page 4 of 4

INVESTIGATION 4 I-CHECK
ELECTROMAGNETIC FORCE

1. The statements below describe how a motor works.

   Write **T** if the sentence is true; write **F** if the sentence is false.

   _____ Electric current travels into the motor through long wires that connect directly to a set of electromagnets.

   _____ The commutator is a switch that turns a set of three electromagnets on and off.

   _____ Only two of three electromagnets are turned on at one time.

   _____ The electromagnets that are turned on attract or repel the permanent magnets that are in the casing.

   _____ When the electromagnets rotate, the commutator is in a new position and one section will turn off while another turns on, continuing the motion to keep the shaft turning.

2. The picture shows two lightbulbs powered by a solar cell. Describe the energy transfers that make the bulbs light, starting with energy from the Sun.

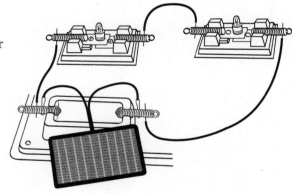

_____

_____

_____

_____

_____

_____

_____

FOSS Next Generation
© The Regents of the University of California
Can be duplicated for classroom or workshop use.

Electromagnetic Force Course
Investigation 4 I-Check
Page 1 of 2

INVESTIGATION 4 I-CHECK
ELECTROMAGNETIC FORCE
. . . .

3. Write **M** next to the statements that describe an electric motor only; write **G** next to statements that describe a generator only; write **MG** next to statements that describe both.

    _____   Includes electromagnets

    _____   Transfers energy from electrical energy to kinetic energy

    _____   Includes a commutator

    _____   Includes a rotor

    _____   Includes permanent magnets

    _____   Transfers energy from kinetic energy to electrical energy

4. Write **R** next to each resource that is renewable; write **N** next to each resource that is nonrenewable.

    _____   The Sun

    _____   Gasoline

    _____   D-cell (battery)

    _____   Wind power

    _____   Geothermal power

    _____   Hydroelectric power

    _____   Natural gas

5. Complete the statement below, using three words from the word bank.

Word bank:	kinetic	transferred	potential
	destroyed	rotate	created

    Energy cannot be _____ or _____; energy can

    only be _____.

FOSS Next Generation
© The Regents of the University of California
Can be duplicated for classroom or workshop use.

Electromagnetic Force Course
Investigation 4 I-Check
Page 2 of 2

Name _____

Date _____ Class _____

1. Fill in the blanks to complete the sentences.

   A _____ is needed to make a motionless object move.

   To make an object in motion stop, you need to _____

   _____.

2. Air hockey is a game played on a specially designed table. Each player tries to slide a puck across the table into a goal. The table has tiny holes that create a cushion of air, lifting the puck a small distance above the playing surface.

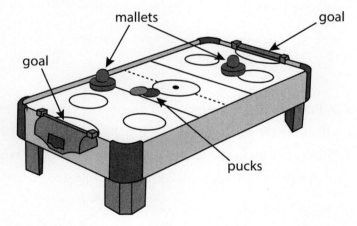

   The cushion of air increases the speed of play because it _____.

   (Mark the one best answer.)

   ○ **A** increases the friction so the puck can be directed more efficiently

   ○ **B** increases the friction to give the puck an additional push

   ○ **C** lowers the friction between the mallet and the puck

   ○ **D** lowers the friction between the table and the puck

FOSS Next Generation
© The Regents of the University of California
Can be duplicated for classroom or workshop use.

Electromagnetic Force Course
Posttest
Page 1 of 7

3. A student pushes with 650 N of force on a cart against a wall. The cart doesn't move. How do you explain the forces at work and why the cart doesn't move?

_____

_____

_____

_____

4. Two teams, Red and Blue, are playing soccer. Three players run after the ball and their feet make contact with the ball at the same time. Student J kicks the ball with 1200 N of force toward the Blue team's goal. Student K kicks the ball in the opposite direction with 550 N. Student L kicks the ball in the same direction as Student K with 650 N of force.

What happens to the ball and why?

(Mark the one best answer.)

 ◯ **A** The ball moves toward the Blue goal because the force from Student K added to the force of Student L is more than the force from Student J.

 ◯ **B** The ball moves toward the Red goal because two students are kicking the ball in the same direction.

 ◯ **C** The ball begins to move toward the Red goal because the force of Student K multiplied by the force of Student L is greater than the force of Student J.

 ◯ **D** The ball stays right where it is because the net force is 0 N.

FOSS Next Generation
© The Regents of the University of California
Can be duplicated for classroom or workshop use.

Electromagnetic Force Course
Posttest
Page 2 of 7

POSTTEST
ELECTROMAGNETIC FORCE

5. Three bar magnets are held together. What are the magnets going to look like when the hands let go of the magnets?

   (Mark the one best answer.)

   ○ **A**

   ○ **B**

   ○ **C**

   ○ **D**

6. A student has two magnets. One magnet is on top of the table. The other magnet is in his hand (1). He puts his hand with the magnet under the table, and the magnet on top of the table begins to move (2). This demonstration provides evidence that _____.

   (Mark the one best answer.)

   ○ **A** the force of magnetism (the magnetic field) goes through some materials

   ○ **B** magnets sometimes repel each other

   ○ **C** magnets sometimes attract each other

   ○ **D** the bottom magnet makes the table into a temporary magnet; which then attracts the magnet on top so it moves

FOSS Next Generation
© The Regents of the University of California
Can be duplicated for classroom or workshop use.

Electromagnetic Force Course
Posttest
Page 3 of 7

POSTTEST
ELECTROMAGNETIC FORCE
• •

7. A student exploring magnets and a bag of other objects discovered that when she touched a nail to a magnet, the nail picked up a paper clip and the paper clip picked up a screw.

Explain to this student why this happens.

_____

_____

_____

_____

_____

_____

_____

Magnets and spacers here   Washers here

8. A group of students conducted an experiment to find out what happens to the force of magnetism between two magnets as the distance between them increases. They used the equipment you see in the illustration. They placed spacers between the magnet in the cup and the magnet on the post to increase the distance between them. They used washers as units to measure how much force it took to pull the two magnets apart.

What conclusion can the students draw from this experiment?

Distance between magnets	Number of washers to pull magnets apart
1 spacer	20
2 spacers	12
3 spacers	8
4 spacers	6
5 spacers	3

_____

_____

_____

_____

_____

FOSS Next Generation
© The Regents of the University of California
Can be duplicated for classroom or workshop use.

Electromagnetic Force Course
Posttest
Page 4 of 7

9. Write **YES** in the box if the circuit will work. Write **NO** in the box if the circuit will not work.

A [   ]   B [   ]

C [   ]   D [   ]

10. Which action would make an electromagnet stronger (able to pick up more paper clips)?

    (Mark the one best answer.)

    ○ **A** Use thinner wire.

    ○ **B** Add another D-cell into the circuit in series.

    ○ **C** Wind fewer wraps around the core (rivet).

    ○ **D** Wind the wraps loosely rather than tightly around the core.

FOSS Next Generation
© The Regents of the University of California
Can be duplicated for classroom or workshop use.

Electromagnetic Force Course
Posttest
Page 5 of 7

11. A group of students removed the plastic insulation around a wire. They wrapped the uninsulated wire around a steel rivet 50 times in order to make an electromagnet. They connected the wires to a battery to make a complete circuit. They tried to pick up some steel washers, but nothing happened. Below is the advice they received from other groups in the class. What advice should they take to get their electromagnet to pick something up?

(Mark the one best answer.)

○ **A** Wrap the wire around the nail more times.

○ **B** Use insulated wire.

○ **C** Add another D-cell to the circuit.

○ **D** Try picking up aluminum paper clips instead of the steel washers.

12. Two groups of students are doing controlled experiments with electromagnets. They each start with an electromagnet that has 30 winds on the core, one D-cell, and a switch. All of the equipment that the groups are using is the same.

Group 1 decides to improve the electromagnet design by changing the number of D-cells in the circuit. They will test 2 D-cells to see if performance is improved.

Group 2 decides to improve the electromagnet design by adding 20 more winds and doubling the thickness of the wire used.

Which group has a better experimental design? Why do you think so?

_____

_____

_____

_____

_____

_____

_____

FOSS Next Generation
© The Regents of the University of California
Can be duplicated for classroom or workshop use.

Electromagnetic Force Course
Posttest
Page 6 of 7

13. Fill in the blanks to complete the sentences.

A generator turns _____ energy into _____ energy.

A motor turns _____ energy into _____ energy.

14. The source of electricity for one student's home is a hydroelectric dam. The student plugs in a lamp at home, and the light goes on. Explain the energy transfers from the water at the dam to the lamp at the house.

_____

_____

_____

_____

_____

_____

_____

_____

_____

_____

_____

_____

_____

_____

_____

FOSS Next Generation
© The Regents of the University of California
Can be duplicated for classroom or workshop use.

Electromagnetic Force Course
Posttest
Page 7 of 7

Notebook Answers

Pushes and Pulls A

Part 1: Pushing and pulling different masses

You will need one spring scale, three loads, a rubber band, and a plastic mat. Use a rubber band to bind three loads together before pulling them.

1. How much force in newtons does it take to **push** these masses?

	Trial 1	Trial 2	Trial 3	Average
1 load				
3 loads		Answers will vary.		

Predict the force required to push 2 loads. _____

What force was needed to push 2 loads? _____

2. How much force in newtons does it take to **pull** these masses?

	Trial 1	Trial 2	Trial 3	Average
1 load				
3 loads		Answers will vary.		

Predict the force required to pull 2 loads. _____

What force was needed to pull 2 loads? _____

3. What is the relationship between the mass of an object and the force needed to slide it across a surface?

The greater the mass, the greater the force required.

FOSS Next Generation
© The Regents of the University of California
Can be duplicated for classroom or workshop use.

Electromagnetic Force Course
Investigation 1: What Is Force?
No. 1—Notebook Master

Pushes and Pulls B

Part 2: Push against push

You will need two spring scales.

4. What happens when spring scale A and spring scale B **both push** with a 4 N force on each other?

Spring scale A 4 N →   ← 4 N **Spring scale B**

The scales push closer together.

5. If you hold spring scale A still and push with a 4 N force with spring scale B, predict what the scale reading on spring scale A will be.
   Prediction: _____ N

A **Hold still**   **B** ← 4 N

a. Measured force on spring scale A: __4 N__
   Was your prediction correct?

b. Explain why that happened.

Spring scale A can't move, but it is being pushed. It pushes back with the same force.

FOSS Next Generation
© The Regents of the University of California
Can be duplicated for classroom or workshop use.

Electromagnetic Force Course
Investigation 1: What Is Force?
No. 2—Notebook Master

Friction Comparison

Surface we are testing: _____

Prediction: Do you think this surface will have more or less friction with the loads than the plastic mat did?

1. How much force does it take to **push** these masses?

	Trial 1	Trial 2	Trial 3	Average
1 load				
3 loads				

2. How much force does it take to **pull** these masses?
(Use a rubber band to bind three loads together before pulling them.)

	Trial 1	Trial 2	Trial 3	Average
1 load				
3 loads				

3. Compare your data with your results from the plastic mat. Does this surface have more or less friction with the loads? Explain your claim using evidence.

> Answers will vary.

FOSS Next Generation
© The Regents of the University of California
Can be duplicated for classroom or workshop use.

Electromagnetic Force Course
Investigation 1: What Is Force?
No. 4—Notebook Master

Pushes and Pulls C

Part 3: Forces on cars

You will need two spring scales and one car.

6. What happens when spring scale A **pushes** with a 2 N force on one side of the car and spring scale B **pushes** with a 3 N force on the other side of the car?

Spring scale A

2 N ⟶

Spring scale B

3 N ⟶

You can't do it. The car rolls to the left, and spring scale A also goes to 3 N if you push on it with 3 N from spring scale B.

7. What happens when A **pulls** with a 5 N force on one side of the car and B **pulls** with a 5 N force on the opposite side of a car?

A

⟵ 5 N

B

5 N ⟶

Nothing moves. The force pulling the car in each direction is equal.

8. Apply a 2 N **pull** with A and a 2 N **push** with B on the car.

A

⟵ 2 N

B

⟶ 2 N

a. Explain what happens to the car when forces are applied.

The car rolls to the left. It is being pushed and pulled to the left.

b. How could you use one pusher to achieve the same result?

You could pull with 4 N or push with 4 N.

9. What causes an object to start moving?

A force exerted on an object makes it start to move.

FOSS Next Generation
© The Regents of the University of California
Can be duplicated for classroom or workshop use.

Electromagnetic Force Course
Investigation 1: What Is Force?
No. 3—Notebook Master

Forces on Carts A

1. In class, students learned that a cart will move when a force of 50 N or greater is applied. A student pushed on the cart with a force of 10 N. The cart did not move. Why not?

   10 N is not enough force to overcome the force of friction.

A

2. Student A pushed on the cart with a force of 500 N. Student B pushed on the other side of the cart. The cart did not move. How much force did student B apply? Explain how you know.

   Student A pushed with 500 N. The cart did not move, so the net force was zero, meaning the force applied on each side was equal.

A   B

3. Student A pushed on the cart with a force of 400 N. Student C pushed on the other side of the cart. The cart rolled to the left. Did student C apply more, less, or the same force as student A? Explain how you know.

   Student C used more than 400 N. Her force exceeded student A's because the net force moved the cart toward student A.

A   C

FOSS Next Generation
© The Regents of the University of California
Can be duplicated for classroom or workshop use.

Electromagnetic Force Course
Investigation 1: What Is Force?
No. 5—Notebook Master

Forces on Carts B

4. Student B pushed on a cart against the wall with 500 N of force. The cart didn't move. How do you explain what happened?

   The net force was zero. This means that the wall was also pushing back with 500 N of force.

B

5. Student C pushed on the cart with 400 N. Student A pushed on the cart with 1,000 N of force. Student B pulled on a rope attached to the cart with 500 N of force. What will happen to the cart and why?

   The cart will move to the left. Student C and student B combined applied 900 N of force to the right, but student A had more than that to the left. The net force is 100 N to the left.

C   A   B

FOSS Next Generation
© The Regents of the University of California
Can be duplicated for classroom or workshop use.

Electromagnetic Force Course
Investigation 1: What Is Force?
No. 6—Notebook Master

Response Sheet—Investigation 2

If you put something like a piece of cardboard between a magnet and an iron nail, the magnet still holds the nail in place, even though the magnet is not touching the nail.

Explain how that happens. Use the word *induce* in your response.

A magnet has an invisible magnetic field extending around it. This field exerts a force on magnetic objects within the field, even if something is between them, like cardboard. The magnet induces temporary magnetism in the iron nail, and the nail and magnet attract each other.

FOSS Next Generation
© The Regents of the University of California
Can be duplicated for classroom or workshop use.

Electromagnetic Force Course
Investigation 2: The Force of Magnetism
No. 8—Notebook Master

Response Sheet—Investigation 1

A dog-sled driver prepared for her race. Her sled was loaded with equipment, but on snow and ice, the dogs moved the heavy sled. Each dog pulled with a force of 600 N.

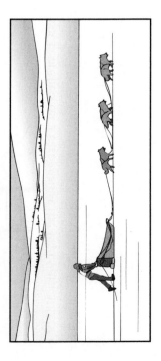

During the race, she came to an area of grass and dirt instead of snow. The sled slowed to a stop. The driver said,

There must be something wrong with one of the dogs. One of them is pulling with a force of less than 600 N.

How would you respond to the dog-sled driver?

It could be true that a dog is pulling with less force. But also the sled is now on a new surface, and the force of friction is different. Grass and dirt would provide more frictional force than snow or ice, so even if the dogs are pulling with the same force, it is not enough to have a net force in the direction of motion.

FOSS Next Generation
© The Regents of the University of California
Can be duplicated for classroom or workshop use.

Electromagnetic Force Course
Investigation 1: What Is Force?
No. 7—Notebook Master

Adding Magnets

1. Position a paper clip on the diagram below so the end is at the 0 mm mark on the ruler. Stand a magnet on edge at 50 mm.

2. Slowly slide the magnet to the left until the paper clip jumps toward the magnet.

3. Record the position of the magnet (jump distance) when the paper clip moved.

4. Collect data for three trials, then repeat with two magnets, three magnets, and four magnets.

Millimeters (mm)

5. Set up a data table for multiple trials with one to four magnets.

Magnets	Trial 1 jump distance (mm)	Trial 2 jump distance (mm)	Trial 3 jump distance (mm)	Average jump distance (mm)
1				
2				
3				
4				

Conclusion

The jump distance increased a lot between the first and second magnet, then just a little more with the third and fourth magnets.

FOSS Next Generation
© The Regents of the University of California
Can be duplicated for classroom or workshop use.

Electromagnetic Force Course
Investigation 2: The Force of Magnetism
No. 9—Notebook Master

Lighting Lightbulbs

Write a prediction for each circuit in the small box. If you think it will light, write "yes." If you think it won't light, write "no."

1. no light

2. no light

3. no light

4. light

5. light

6. no light

FOSS Next Generation
© The Regents of the University of California
Can be duplicated for classroom or workshop use.

Electromagnetic Force Course
Investigation 3: Electromagnetism
No. 10—Notebook Master

Testing Electromagnet Variables

Variable we are testing: _____

Standard electromagnet—number of paper clips lifted			
Trial 1	Trial 2	Trial 3	Average

Modified electromagnet—number of paper clips lifted			
Trial 1	Trial 2	Trial 3	Average

Data Analysis

1. Which device was more effective, your standard or modified electromagnet? Use evidence to support your response.

Answers will vary.

Class Analysis

2. Looking at the class data, which modifications are the most effective? Use evidence to support your response.

The most effective modifications are more wraps of wire, thicker wire, and more batteries in series. Each of these things made the electromagnet stronger.

3. Answer on a separate page: If you built an electromagnet with multiple modifications, how would you build the most powerful electromagnet?

Answers will vary.

FOSS Next Generation
© The Regents of the University of California
Can be duplicated for classroom or workshop use.

Electromagnetic Force Course
Investigation 3: Electromagnetism
No. 12—Notebook Master

Response Sheet—Investigation 3

Student 1 said,

Magnetic force and electric charges are similar, because opposites attract and likes repel.

Student 2 said,

Magnetism and electricity are similar because they both have a force field extending out from them so they don't have to touch to work.

How would you respond to these students?

To student 1, I would say that I agree they are similar in this way, although there are other differences between magnetic force and electric charges like the materials they can or cannot pass through.

To student 2, I would say that they are different in this way. Electricity can only move through conductive materials, not insulators like plastic. Magnetic fields extend through materials like plastic.

FOSS Next Generation
© The Regents of the University of California
Can be duplicated for classroom or workshop use.

Electromagnetic Force Course
Investigation 3: Electromagnetism
No. 11—Notebook Master

Energy Transfers

1. What energy transfers are happening to make the motor run?

   Kinetic energy from your hand transfers to make the motor run. The kinetic energy of the shaft transfers energy to the electromagnet coils, generating electric current. The electrical energy travels to the motor, turning on electromagnets, which make the shaft rotate (kinetic energy).

2. Compare the energy transfers of a motor to the energy transfers of a generator.

   The generator basically transfers energy in the opposite pathway as the motor (from kinetic energy when your hand cranks to kinetic energy of the rotor to electrical energy in the electromagnets which travels through the circuit).

3. Compare the inner components of a battery/motor system to the inner components of a hand-crank generator.

   Both have a shaft connected to an electromagnet within a case.

4. What do you think will happen when we hook up batteries to the wires of the hand-crank generator?

   If it has the same internal components and overall design as a motor, batteries might make it run like a motor.

5. What happened when you hooked up batteries to the hand-crank generator in class?

   The generator ran as a motor. The crank turned. The energy source for its kinetic energy was the batteries.

FOSS Next Generation
© The Regents of the University of California
Can be duplicated for classroom or workshop use.

Electromagnetic Force Course
Investigation 4: Energy transfer
No. 13—Notebook Master